Twenty-first Century Perspectives on War, Peace, and Human Conflict

Series Editors
Christine Cheng
King's College London
London, UK

John Karlsrud
Norwegian Institute of International Affairs
Oslo, Norway

This Palgrave Macmillan book series invites methodologically pluralist and interdisciplinary approaches to the study of peace and conflict. We aim to bring new knowledge and pathways for understanding conflicts and conflict actors. Thematically, we welcome monographs, strong edited works, and handbooks on peacekeeping, peacebuilding, trauma, war-to-peace transitions, statebuilding, violent extremism, corruption, non-state armed groups, global and regional organisations, and inter-organisationalism relating to conflict management.

We are particularly keen on interdisciplinary work– especially where politics and international relations intersect with sociology, anthropology, law, psychology, geography, criminology, technology, gender studies, and area studies. And we are broadly interested in conflict writ large, beyond the bounds of civil and interstate war, stretching over into urban violence, sexual violence, post-colonial reparations, transitional justice, etc. Our goal is not just for your manuscript to be published, but to be read, discussed, contemplated, and acted upon. To that end, we seek research findings that are compelling, and writing that is memorable and immersive.

The principal aim of this series is to provide educators and students, decision-makers, and everyday citizens with contemporary, cutting-edge thinking about the roots of conflict, international responses, and the conditions for just and enduring peace. The series includes unorthodox and cross-disciplinary approaches to these topics, as well as more traditional social scientific and humanistic monographs.

We strongly encourage early-career scholars and innovative researchers – especially those from under-represented backgrounds – to submit manuscripts for review. All titles in the series are peer-reviewed and we aim to provide rapid and constructive feedback. We also welcome open access arrangements.

For an informal discussion for a book in the series, please contact the series editors Christine Cheng (christine.cheng@kcl.ac.uk) and John Karlsrud (jka@nupi.no).

For the correct copy of Palgrave's book proposal form, please contact Palgrave editor Isobel Cowper-Coles, Editor for International Studies, isobel.cowpercoles@palgrave.com

Arthur Boutellis

The UN Stabilization Mission in Mali (MINUSMA)

Peacekeeping Caught in the Geopolitical Crossfire

palgrave
macmillan

Arthur Boutellis
International Peace Institute
New York, NY, USA

ISSN 2945-6053 ISSN 2945-6061 (electronic)
Twenty-first Century Perspectives on War, Peace, and Human Conflict
ISBN 978-3-031-66465-6 ISBN 978-3-031-66466-3 (eBook)
https://doi.org/10.1007/978-3-031-66466-3

This Palgrave Macmillan imprint is published by the registered company Springer Nature Switzerland AG.
The registered company address is: Gewerbestrasse 11, 6330 Cham, Switzerland

Preface

From 2013 through 2015, I served as a United Nations staff with the United Nations Multidimensional Integrated Stabilization Mission in Mali (MINUSMA), on the team that supported the inter-Malian peace negotiations that took place in Algiers. This was my last of a series of field assignments with the UN since 2005. In January 2015 armed clashes were taking place between two coalitions of armed groups in northern Mali (the *Coordination of Azawad Movements*—or CMA—rebel groups on one side, and the *Platform* supported by the government of Mali on the other) that were attempting to make territorial gains they could leverage prior to the next round of peace negotiations that were due to resume in Algiers in February 2015. On January 20, CMA rebels were firing heavy machine guns and 122 mm rockets toward the town of Tabankort which was occupied by *Platform* groups. A few days earlier, MINUSMA peacekeepers on a long-range patrol had stopped and established a position in Tabankort to try to prevent clashes between the two coalitions of armed groups and to protect the civilian population in the area. But MINUSMA had become caught in the crossfire.

As 122 mm rockets fired by the CMA rebels were getting closer to the peacekeepers' position, MINUSMA contacted the political and military leadership of the rebel groups to urge them to stop but they didn't. Considering the imminent danger to peacekeepers, MINUSMA decided to send two of its Dutch-operated Apache military attack helicopters to the area to monitor the situation and to deter further firing. As CMA rebels continued to fire, MINUSMA helicopters first fired warning shots but that also failed to halt the shooting. After careful review of the UN mission's mandate (which included the *Protection of civilians and United Nations personnel*) and rules of engagement, MINUSMA helicopters fired hellfire missiles and destroyed a pickup truck from which rockets were fired, killing, and injuring a few CMA rebels.

MINUSMA's robust response in Tabankort triggered a series of demonstrations the next day against the UN mission in the rebel stronghold of Kidal, which resulted in the vandalization of the airstrip rendered inoperable for months. MINUSMA subsequently facilitated the negotiation of a local ceasefire in Tabankort to prevent further violence and obtain the withdrawal of both CMA and *Platform* fighters, which this time triggered violent protests in front of the MINUSMA camp in Gao by pro-*Platform* groups. Despite these challenges, the fifth and final round of inter-Malian negotiations took place in Algiers from 16 February to 1 March, which opened the way for the signing of the 2015 peace agreement in May-June by the government of Mali, the *Platform* armed groups and finally the CMA armed groups.

Ten years after I first arrived in Mali, as I am writing this book, the Mali transition government, on June 16, 2023, formally asked for the "withdrawal, without delay" of the decade-old mission. MINUSMA is once again caught in the crossfire, but this time of competing geopolitical interests of Western powers—and especially France as the *penholder* on Mali in the Security Council—on the one side, and on the other side the host country led by a military junta supported by Russia (and the *Wagner Group* mercenaries which arrived in Mali late 2021) defying the Security Council, in a context of return of East-West rivalries exacerbated by the 2022 Ukraine invasion. The Security Council subsequently decided to "terminate MINUSMA's mandate as of June 30, 2023" and that the UN mission would withdraw by the end of 2023.

This final MINUSMA episode has been a striking illustration of how evolving power rivalries have increasingly played out at the UN in general in recent years, including around its most visible tool which is UN peacekeeping. The Malian abrupt decision made MINUSMA's rushed withdrawal during the second half of 2023 even more challenging, as fighting between the Malian army and northern rebel groups resumed after years of relative stability. On January 25, 2024 the Malian transition authorities declared the end of the 2015 Inter-Malian Agreement for Peace and Reconciliation, further closing the MINUSMA chapter since the peace agreement had been the UN mission's main strategic priority and *raison d'être* over the past few years and considered by the "international community" as essential to stabilizing the country.

Having been a privileged actor and observer of how the changing global order and power rivalries have increasingly affected the UN both in New York and in field operations over the past decade or so,—MINUSMA being the most rich and iconic laboratory of contemporary UN peace operations, I hope this book will first of all tell an honest, informed and analytical story of the UN multidimensional stabilization operation in Mali, for the record. But the ambition of the book is also through the story of MINUSMA to tell a bigger story of a major turning point for UN peacekeeping—and notably its large multidimensional stabilization missions—the reasons behind it and the implications for the future of UN peace operations more generally.

Peacekeeping is witnessing the end of three decades of expansion made possible by the post-Cold War international liberal order and the adaptation and complexification of the peacekeeping tool to deal with intra state wars. Until recently, UN peace operations had been largely shielded from power rivalries and Security Council divisions, generally between the US, France and the UK on one side (the "P3") and Russia and China on the other side (the "P2"), over issues such as Syria or Ukraine. Members of the Council had indeed demonstrated a unique ability to "compartmentalize" their differences, and thus maintain a common front on the deployment of peace operations in Africa notably. But UN peace operations have now become yet another, albeit still mostly indirect, field of confrontation between major and middle powers in a context of 'crisis of multilateralism' and contestation of the 'liberal order' perceived by a growing number of member states as a 'Western order.'

As both a practitioner and an academic, and a former MINUSMA staff who has visited Mali at least once a year since from 2015 to 2022, I hope to bring through this book unique insights from my privileged access both in New York and in Mali, long-term participant observation, enriched with interviews with diplomats and UN officials in New York, in certain capitals, and in field operations. I also used primary sources, notably minutes of Security Council meetings, and secondary sources such as think tank and press reports. The book therefore offers a hybrid empirical-theoretical perspective. While MINUSMA has been the subject of many think tanks reports and journal articles, this is the first book dedicated to the UN mission in Mali.

Like many other "experts," I had until then mainly focused on practical or "micro" policy questions aimed at improving the functioning of the UN, and particularly its peace operations, without really making the link with theoretical debates at a more "macro" level, to use the terminology of Roland Paris. Through this book, I also hope to contribute to the rapprochement between peace operations and international relations, by promoting an original "realist-constructivist" analytical framework, as an alternative to the dominant liberal approach when it comes to the study of UN peacekeeping. While the study of peace operations cannot of course be reduced to that of geostrategic rivalries, this book aims to demonstrate that we cannot understand recent (as well as historical) developments in UN peace operations without better factoring in these complex power rivalries and their rapid evolutions.

New York, NY, USA J. Arthur Boutellis

Acknowledgements

First, I would like to express my deep thanks to all the people who, over the past few years, have guided and enriched my work of research, conceptualization, theoretical development, and writing, starting with my doctoral thesis supervisor, Jean-Vincent Holeindre. I would also like to thank the members of my PhD jury thesis, namely professors Guillaume Devin, Delphine Allès, Frédéric Charillon, Ronald Hatto, Lise Morjé Howard, and Marie-Joëlle Zahar, as well as the members of the Léon Bourgeois Prize jury for recognizing the contribution of my PhD thesis to academic work on international cooperation.

I thank the diplomats, researchers, UN employees and former colleagues who agreed to share with me their experiences and expertise as part of my research, in New York, in Paris and in particular at the Thucydides Center of Panthéon-Assas University, and on the ground in UN peace operations, in Mali in particular. They will recognize themselves. I would of course thank my family for their understanding (and misunderstandings) while I undertook this research and writing work "on top of everything else."

I particularly thank Séverine Autesserre, Camille Bayet, Matts Berdal, Nicolas Degrais, Lise Howard, John Karlsrud, Youssef Mahmoud, Delphine Mechoulan, Alexandra Novosseloff, Namie Di Razza for their enlightened advice and encouragement at the beginning of my work and for having helped me forge the conviction that I could modestly contribute to advancing the reflection on UN peace operations in the field of international relations.

I express my gratitude to those who took the time to read different excerpts and versions of my manuscript and allowed me to improve, and in certain cases correct, the content: Camille Bayet, Mohamed Bouabdallah, Alexandre Carette, Emmanuelle Cousin, Namie Di Razza, Marie Forestier, Christophe Gargot, Adeline Guillemain, Naomi Miyashita, Colombe de Nicolay, and Alexandra Novosseloff.

Last but not least, special thanks to the co-editors of the series Christine Cheng and John Karlsrud, as well as to Isobel Cowper Coles and Raju Heggidi at Palgrave Macmillan for making this book possible.

Contents

About the Author

J. Arthur Boutellis is a Non-Resident Senior Adviser at the *International Peace Institute* (IPI) in New York, where he was Director of the Brian Urquhart Center for Peace Operations from 2015 to 2017, and a Senior Advisor with *TrustWorks Global.* He teaches at graduate-level at *Columbia University* and *Sciences Po Paris.* Arthur previously served in four UN peace operations including in Mali (MINUSMA) where he supported the 2014–15 Mali peace negotiations as part of the UN Mediation Team. He previously worked with humanitarian NGOs and think tanks focused on the Middle East and Africa. He holds a master's degree in public affairs from Princeton University, a PhD from *Université Paris Panthéon-Assas* and is a 2020 auditeur of *the Institut des Hautes Études de Défense Nationale (IHEDN)*. He serves on the Editorial Board of *International Peacekeeping* and a member of both the *Observatoire du multilatéralisme et des organisations internationales* (GRAM) and the Global Initiative against Transnational Organized Crime (GI-TOC). A list of his publications is available here: https://www.ipinst.org/by/arthur-boutellis.

Abbreviations

A4P	Action for Peacekeeping
A4P+	Action for Peacekeeping Plus
AFISMA	African-led International Support Mission to Mali
AQIM	Al-Qaida in the Islamic Maghreb
ASIFU	All Source Information Fusion Unit
AU	African Union
BATFAR	*Bataillon des Forces Armées Reconstituées*
C-34	Special Committee on Peacekeeping Operations
CIMIC	Civil-Military Coordination
CJA	Congress for Justice in Azawad
CMA	Coordination of Azawad Movements (armed group coalition)
CMFPR	Coordination of Patriotic Movements and Forces of the Resistance
CMI	Coordination des mouvements de l'inclusivité
CNDRR	National DDR Commission
CPA	Coalition of the People for Azawad
CRSS	Security Sector Reform Commission
CSA	Follow-Up Committee (to the 2015 inter-Malian peace agreement)
CSE	Monitoring and Evaluation Committee to the 2013 Ouagadougou agreement
CTMS	Joint Technical Commission for Security (following Ouagadougou agreement)
CTS	Technical Security Commission (following the 2015 peace agreement)
CVR	Community Violence Reduction
DDR	Disarmament, demobilization, and reintegration

DRC	Democratic Republic of the Congo
ECOWAS	Economic Community of West African States
EU	European Union
EUCAP	European Union civilian crisis management mission in Mali
FC-G5S	G5 Sahel Joint Force
FLM	Macina Liberation Front
GATIA	Self-Defense Group of Imghad Tuaregs and Allies
HCUA	High Council for the Unity of Azawad
ISR	Intelligence, Surveillance and Reconnaissance
JNIM	Jama'at Nusrat al-Islam wa al-Muslimeen (coalition of al-Qaeda-groups)
MAA	Arab Movement of Azawad
MDSF	Malian Defence and Security Forces
MINUSCA	United Nations Multidimensional Integrated Stabilization Mission in the Central African Republic
MINUSMA	United Nations Multidimensional Integrated Stabilization Mission in Mali
MONUSCO	United Nations Organization Stabilization Mission in the Democratic Republic of the Congo
MNLA	National Movement for the Liberation of Azawad
MOC	Operational Coordination Mechanism (for mixed patrols)
MPSA	Popular Movement for the Salvation of Azawad
MSA	Movement for the Salvation of Azawad
NIC	National Integration Commission
PoC	Protection of Civilians
PRV	Project for the Reduction of Violence in Central Mali
QIPS	Quick Impact Projects
SAGE	Situational Awareness Geospatial Enterprise
SOFA	Status of Forces Agreement
SOMA	Status of Mission Agreement
SOP	Standard Operating Procedure
SRSG	Special Representative to the Secretary-General
TCC	Troop Contributing Country
UNCT	United Nations Country Team
UNMISS	United Nations Mission in the Republic of South Sudan
UNOWAS	United Nations Office for West Africa and the Sahel
UNPOL	United Nations Police
UNRC	United Nations Resident Coordinator

List of Figures

CHAPTER 1

Introduction

"Too often there is a disconnect between leadership and power. We see remarkable examples of leadership; but they are not generally associated with power. And power is not always associated with the necessary leadership."[1]
Antonio Guterres, September 22, 2020

On June 16, 2023, the representative of Mali stunned the United Nations Security Council when it asked for the "withdrawal, without delay" of the decade old United Nations Multidimensional Integrated Stabilization Mission in Mali (MINUSMA). Although relations between the Malian transitional government and the UN mission had deteriorated, the timing of the demand still caught by surprise the UN Secretary-General's Special Representative and Head of MINUSMA El-Ghassim Wane as well as most Council members. Some diplomats briefly considered options for keeping MINUSMA in place, but quickly the Security Council resigned itself to the fact that the host country's consent to the UN operation had been withdrawn and that the best that could be negotiated was a swift but orderly exit of MINUSMA's civilian and military personnel by the end of December 2023.

[1] António Guterres, United Nations Secretary-General, addresses the general debate of the 75th Session of the General Assembly of the UN (New York, September 22–29, 2020), https://www.youtube.com/watch?v=V2ZR45Bs3gs.

A. Boutellis, *The UN Stabilization Mission in Mali (MINUSMA)*, Twenty-first Century Perspectives on War, Peace, and Human Conflict, https://doi.org/10.1007/978-3-031-66466-3_1

In a changing global order, MINUSMA has in many ways been the most rich and iconic laboratory of contemporary UN peace operations. It was one of the largest "stabilization" missions (with an annual budget of $1.2bn), with significant European and Chinese contingents working with a majority of African peacekeepers, operating in parallel to a French force—and later Russian *Wagner* mercenaries—in an environment affected by terrorism, which also meant that it has also been the deadliest of the UN missions with over 300 casualties. MINUSMA ultimately became the collateral damage of new geopolitical rivalries heightened by the growing influence of Russia on the African continent—supported by deployments of *Wagner Group* mercenaries and Russian instructors—and the February 2022 invasion of Ukraine.

While the analysis of UN peace operations cannot be reduced to that of geostrategic rivalries only, this book which hopes to make both an empirical and theoretical contribution to the field, demonstrates that it is not possible to understand historical developments in UN peace operations without factoring the evolution of these power rivalries. In that sense the book offers a realist-constructivist approach that aims to challenge the dominant liberal paradigm when it comes to the study of UN peace operations. UN peace operations have taken many forms over the past 75 years, some modest in size and mandate, others far more ambitious and multidimensional. They have adapted and been shaped by changes in the international order in which they have operated, but their future is now called into question under the current context of exacerbated geopolitical rivalries.

The book is centered around the decade-long life of the United Nations Integrated Stabilization Mission in Mali (MINUSMA) divided into three distinct phases during which the UN mission experiences the evolution of power rivalries between a few key states differently. The first phase of the life of MINUSMA (2013–2015) is marked by a relatively unanimous Security Council behind the French penholder on the direction to be given to the UN peace operation in Mali. France even manages to mobilize European troop-contributing countries who are thus returning to the African continent after many years of absence.

The second phase (2016–2019) is marked by growing differences of opinion between the permanent members of the Security Council, including within the P3[2] between France and the United States over the continued relevance of a peacekeeping operation in an environment plagued by the threat of terrorism. The US however accepts to continue supporting the UN mission on condition it puts a greater emphasis on the protection of civilian in Central Mali.

During the third and final phase of the life of MINUSMA (2020–2023) the France-United States divide is replaced by a more classic East-West divide, the result of exacerbated rivalries for power and influence between on the one side France and the US, and on the other side the host state governed by a military junta and supported by Russia and *Wagner group* mercenaries. These new divisions that emerged from a new set of players in Mali itself, were exacerbated by Russia's invasion of Ukraine in February 2022 and quickly polluted Security Council debates in New York.

MINUSMA found itself trying to navigate the different positions taken by Council members and a host state whose transitional authorities sought to consolidate their power at all costs and grew increasingly frustrated with the UN mission on the ground. The Malian transition government, upset with the French pen holder,[3] but also probably encouraged by its new Russian ally, formally requested the departure of the mission on June 16, 2023. The UN mission and UN Secretariat ultimately had little control or influence over these growing geopolitical tensions but got caught in the crossfire.

The last two chapters of the book draw lessons from the MINUSMA experience and the return of power rivalries at the UN for the future UN peace operations in a post-liberal international order. The book demonstrates that peacekeeping operations, which had largely been consensual among UN Member States despite growing divisions within the Security Council—particularly since the NATO bombing campaign that toppled Gaddafi in 2011, the conflict in Syria and the 2014 invasion of Ukraine—have now also become a field of confrontation between major and middle powers with increasingly divergent interests and values, both at the United Nations headquarters in New York and in the field, notably in Africa.

[2] The "P3" commonly designates the US, France, and the UK as generally "like-minded" permanent members of the Security Council; the "P2" designated Russia and China as the other two permanent members of the Security Council; together these form the permanent five of "P5".

[3] Penholding has been defined by the Security Council as "the informal arrangement whereby one or more Council members (as 'penholder(s)') initiate and chair the informal drafting process." UN Doc. S/2017/507, para. 78.

These UN operations are both an instrument and an object of these new power dynamics.

The book also demonstrates that these growing power rivalries, particularly during the more recent period, are beginning to have deleterious consequences, both at headquarters level and for field operations. In New York, these rivalries play out at the Security Council, the General Assembly but also the United Nations Secretariat (or the bureaucracy), through the negotiations of resolutions and declarations, budgets, but also around values and norms, as well as in increasingly politicized senior UN appointments.

In field operations, other power games also manifest themselves through troop contributions, the growing assertiveness of host countries, and various "parallel forces." While this member state influence if often more indirect in the field, which theoretically gives greater maneuver room for UN operations, sometimes conflicting interests of different key powers become more direct and can obstruct the work of a peace operation, as it became the case in Mali, but also in the Central African Republic, Libya or Syria.

Richard Gowan has noted on multiple occasions the ability of Security Council members to "compartmentalize" their differences—arguing bitterly over certain issues like Syria, Libya, Yemen, Myanmar, Crimea, the Democratic People's Republic of Korea, and Venezuela, etc. but still passing resolutions renewing the mandate UN peace operations. This has held true even after the February 2022 invasion of Ukraine.[4] However, over the past decade and particularly since 2018, which we argue marks a turning point, power rivalries have started impacting the course of peace operations more directly and risk making these operations less impartial, less legitimate, and less relevant.

The book asks the question of the future of UN peace operations, in a context of power rivalries exacerbated by the growing importance of China at the UN and the China-US rivalry, Russia's growing influence on the African continent, and the effects of the war in Ukraine. It identifies two main risks: that current rivalries continue to constrain UN operations, and to undermine their impartiality and relevance; or that on the contrary they lead to great and middle powers disengaging durably from peace operations that they would see as neither delivering on the ground, nor really serving their interests. There is however no fatalism in this and

[4] Richard Gowan, "The Price of Order", 2018.

multilateralism is compatible with the exercise of power provided member states are engaged in a constructive manner.

Rather than try to shelter itself and risk irrelevance, the UN Secretariat should embrace power rivalries and better consider power dynamics at headquarters and in field operations without resigning itself to them. For their part, member states would be mistaken to neglect multilateralism in the context of the hardening of international relations, or to focus exclusively on bilateral and ad hoc peace enforcement arrangements. Today's world requires diversified strategies of cooperation, power, and influence, for which UN peace operations remain a useful vehicle.

UN Peace Operations, Shaped by Changes in the Global Order

Since the creation of the United Nations, more than a million peacekeepers have served in more than seventy peace operations. Peace operations were not initially provided for as such by the Charter of the United Nations signed in San Francisco on June 26, 1945. The Charter envisages that "in order to enable the United Nations to take urgent military measures, Members shall hold immediately available national air-force contingents for combined international enforcement action." (Article 45). But the circumstances of the Cold War led member states not to activate this possibility of a military tool on a multilateral scale.

It is this absence of a military tool which ultimately led the UN and the Security Council to improvise and replace it with an ad hoc *system* of peacekeeping operations, the *Blue Helmets*, which resulted from an interpretation of the articles of the Charter and its spirit. UN peace operations have taken many forms over the past 75 years, some modest, others much more ambitious. They adapted to and they were shaped by developments in the international order in which they operated at different periods in history. Peace operations have also become the most visible and high-profile activity of the world Organization and continue to occupy a large part of the agenda of its Security Council.

It is paradoxically when the UN appears paralyzed by divisions that member states assert their power and influence through United Nations peace operations, in a context of renewed tensions and geopolitical rivalries. Contrary to what one might think and what has been written in the academic and political literature, this hardening of international relations does not necessarily translate into more robust or offensive UN operations.[5] The latter remain the prerogative of the national forces of a few great and regional powers, NATO and ad hoc coalitions. These games of power and influence by member states with increasingly divergent interests and values, however threaten the future of UN peace operations which remain one of the main peace and security tools at the disposal of the Security Council and an important vehicle through which member states practice multilateralism.

UN peace operations are understood in this book as encompassing both peacekeeping and "special political missions" without troops. While this distinction is largely bureaucratic and budgetary, it persists to date.[6] Peace operations range from modest observation missions (sometimes described as "traditional" or "ceasefire observation" missions) from the beginnings of peacekeeping in 1948 to more complex so-called multidimensional operations integrating hundreds of civilian experts and thousands of peacekeepers, with large aircraft fleets and even maritime components in some cases. These UN missions sometimes operate in par-

[5] See for example Robert Muggah, "The United Nations Turns to Stabilization," IPI Global Observatory, December 5, 2014; Cedric de Coning, *UN Peacekeeping Doctrine in a New Era: Adapting to Stabilization, Protection and New Threats*, 2017, Routledge; Aditi Gorur, *Defining the Boundaries of UN Stabilization Missions*, Washington, DC: Stimson Center, 2016; John Karlsrud, "The UN at War: Examining the Consequences of Peace-Enforcement Mandates for the UN Peacekeeping Operations in the CAR, the DRC and Mali." *Third World Quarterly* 36, no. 1, 2 January 2015; John Karlsrud, "Towards UN Counter-Terrorism Operations?" *Third World Quarterly* 38, no. 6, June 3, 2017: 1215–31; Charles T. Hunt, "All Necessary Means to What Ends? The Unintended Consequences of the 'Robust Turn' in UN Peace Operations," *International Peacekeeping* 24, no · 1, January 2017, 108–31.

[6] "Special political missions" include political missions and good offices missions. Unlike peacekeeping operations financed by a dedicated budget, they are financed by the regular budget of the UN Secretariat. Special political missions have multiplied over the last twenty years and have taken new and more operational forms. Beyond the budgetary question, what defines these political missions is that they do not have troops or blue helmets and are authorized under Chapter VI of the Charter. But they are increasingly taking on the appearance of peacekeeping operations due to their often relatively large size (even if their budgets remain modest compared to those of large peacekeeping operations) and their "multidimensionality" with sometimes military and police advisors (usually without uniform), and/or "guard units" provided by troop-contributing countries to protect offices in the most dangerous environments, in Iraq, Somalia, Libya and the Republic Central African. Some of the most recent political missions include the 2016 mission in Colombia (renamed verification mission in 2017), the 2019 mission in support of the Hudaydah Agreement in Yemen (UNMHA) and in 2020 the mission in Sudan (UNITAMS).

allel with other international forces or following non-UN military interventions. UN peace operations have professionalized considerably and become more complex over the years but many of these different "models" of peace operations continue to coexist until today.

Over seven decades of existence, the UN has equipped itself with a broad toolbox and range of know-how and instruments it can draw from. Since 2007, the UN has provided logistical and financial support to the African "peace enforcement" operation, the African Union Mission in Somalia (AMISOM)[7] which became the African Union Transition Mission in Somalia (ATMIS) in April 2022. In 2014, the first regional emergency "health" operation of the United Nations (the UN Mission for Ebola Emergency Response or UNMEER) was put in place in response to an epidemic in West Africa. In 2017, the Security Council authorized MINUSMA to provide operational and logistical support to the G5 Sahel joint counterterrorism force.

The closure of MINUSMA at the end of 2023 confirms a trend of peacekeeping contraction, with the closing of major operations—each with more than 10,000 peacekeepers each and an average longevity of 13 years—that started in 2017–2018 in Liberia, Côte d'Ivoire, and Haiti, followed by the end of the hybrid UN-AU operation in Darfur at the end of 2020. And no new peacekeeping operation authorized by the Council since MINUSCA in the Central African Republic in 2014. Following the resumption of fighting between rival militaries in Sudan, the UN Integrated Transitional Assistance Mission in Sudan (UNITAMS)—a sizeable political mission—also closed at the end of 2023, and the United Nations Organization Stabilization Mission in the Democratic Republic of the Congo (MONUSCO) is also due to close in 2024. The annual peacekeeping budget thus fell from more than 8 billion dollars in 2016–2017 to $5.5 billion ($6.1 billion including a six-month allocation of $590 million for the withdrawal of MINUSMA) for the period from July 1, 2023 to June 30, 2024, for the nine remaining peacekeeping operations.

At the same time, a number of "special political missions" (SPM) without a military component (peace operations with a "lighter footprint" and generally much smaller budgets) have been authorized by the Security Council in recent years: The United Nations Verification Mission in Colombia in 2016 (which included 450 unarmed military observers in

[7] Paul D. Williams and Arthur Boutellis, "Partnership Peacekeeping: Challenges and Opportunities in the United Nations-African Union Relationship." African Affairs. 2014.

civilian clothes), the UN Mission to support the Hudaydah Agreement (UNMHA which includes 75 observers also unarmed and in civilian clothes) in 2019, and the UN Integrated Transition Assistance Mission in Sudan (UNITAMS) in 2020, which was meant to provide support to Sudan during its political transition to democratic rule. These are illustrations of an increasingly "tailored" approach.

United Nations peace operations are envisaged in this book in a broad sense as all multinational interventions decided by the Security Council and led by the UN Secretariat or in rare cases the UN General Assembly. These operations can be aimed at preventing conflict or maintaining, restoring, consolidating, or imposing peace, facilitating humanitarian assistance and/or stabilizing a situation through the deployment of military personnel, police and/or civilian under the direction of the UN.

Indeed, one of the particularities of UN peace operations is their politico-military dimension but also to the fact that they are generally led by a civilian head of mission on the ground (unlike NATO or EU military operations) which often has significant room for maneuver in implementing the mandate that the Security Council gives to the mission. There is therefore a "double delegation of authority," from the Security Council to the UN Secretariat and from the Secretariat to the mission on the ground.

These operations are authorized under Chapter VI (Peaceful settlement of disputes) or VII (Action with Respect to Threats to the Peace, Breaches of the Peace, and Acts of Aggression) of the UN Charter. Peacekeeping operations can take multiple forms and evolve depending on the contexts and geopolitical realities, but they are guided by three fundamental principles which are the consent of the parties, impartiality, and the non-use of force except in self-defense or defense of the mandate, including protection of civilians mandates.[8]

These UN peace operations are also the result of a unique sharing of institutional responsibilities. The Security Council gives the mandate to the operation and therefore the strategic direction. The General Assembly decides on the financing of these operations, particularly in the Fifth Committee responsible for administrative and budgetary matters, but also gives them a normative framework in the Special Committee on Peacekeeping Operations (known as "C-34").

[8] Official website of the Department of Peace Operations (DOP), https://peacekeeping.un.org/fr/what-is-peacekeeping.

The United Nations Secretariat (or bureaucracy) implements the decisions of member states. "Contributing countries" provide military contingents and police officers to operations. And we must not forget that governments of the host countries of these operations also have a key role because they formally consent to the operations and can withdraw this consent as we've seen in the case of Mali in June 2023, and peacekeeping operations cannot operate effectively when host-country interlocutors oppose their presence or mandate.[9] UN peace operations are therefore a concerted effort to maintain international peace and security.

The unique strength of these UN peace operations, unlike other peace operations (NATO, regional organizations or ad hoc coalitions for example), lies in their multilateral and often "multidimensional" character and the legitimacy conferred upon them by the United Nations Charter and the wide range of member states that contribute to the peacekeeping operations, politically, financially, and in troops and police officers—with more than 130 troop and police contributing countries (TCC/PCCs).

A widespread idea among liberal internationalists is that "multilateralism implies that partners enter into relationships on the basis of rules rather than on the basis of power"[10] from which states can engage in reciprocity and institutionalized cooperation.[11] But power relations are also exercised through the multilateral system. In this regard, John Ikenberry evokes strong multilateral "institution-building" states, which agree to limit the use of their power, in order to obtain from middle powers and smaller states that they participate and comply with the institutions set up in line with the interests of the strongest.[12] Secretary-General Dag Hammarskjöld, in the midst of the Cold War, declared that "it is not the

[9] Julie Gregory and Lisa Sharland, Host-Country Consent in UN Peacekeeping, The Stimson Center, 2023, Washington, D.C., USA.

[10] Martha Finnemore, "Fights about Rules: The Role of Efficacy and Power in Changing Multilateralism," *Review of International Studies* 31, no · S1 (December 2005): 187–206, https://doi.org/10.1017/S0260210505006856. P195.

[11] Gilford John Ikenberry, *Liberal Leviathan: The Origins, Crisis, and Transformation of the American World Order*, Princeton Studies in International History and Politics (Princeton, NJ Oxford: Princeton University Press, 2011). p. 18.

[12] G. John Ikenberry, *After victory: institutions, strategic restraint, and the rebuilding of order after major wars*, Princeton studies in international history and politics (Princeton: Princeton University Press, 2001).

Soviet Union or any other great power that needs the United Nations for its protection: it is all the others."[13]

This "bargain" between great powers and other member states is nevertheless being called into question today as member states are increasingly divided at the UN on various issues and as great powers (and some regional powers) seem less willing to voluntarily limit their power. More and more UN peace operations have also become a field (among others) of confrontation between powers in a context of questioning of the multilateral system and the "liberal international" order. A growing number of member states, and notably regional powers but also states hosting UN peace operations (Mali asked France to give up the pen on the Mali situation in the Security Council[14]), indeed call into question the representativeness and legitimacy of institutions such as the Security Council reflecting post-World War II circumstances.

The Contemporary Study of Peace Operations

It is interesting to note from the outset that contemporary literature on peace operations remains largely empirical and does not really fit squarely into the theoretical fields of political science and international relations, or indeed into that of security studies, peace studies, strategic studies, or war studies. One of the difficulties is that UN peace operations are truly at the intersection of several disciplines and thus find themselves under-studied by each of these disciplines. The observation made by Roland Paris twenty years ago, that the study of peace operations has suffered from a "cult of political relevance," remains largely true today on a global scale:

> Even though the study of peace operations is booming, it remains largely cut off from the rest of political science. Contributors to this literature have primarily focused on practical policy-related issues, such as the design and conduct of particular missions, rather than building bridges between the study of peace operations and broader theoretical debates in the discipline.[15]

[13] Dag Hammarskjöld, Statement before the General Assembly, October 3, 1960 (A/PV. 883).

[14] RFI, "ONU: Bamako ne veut plus de la France comme "plume" des résolutions sur le Mali au Conseil de sécurité", March 4, 2023.

[15] Roland Paris, "Broadening the Study of Peace Operations," *International Studies Review* 2, no · 3 (December 2000): 27–44, https://doi.org/10.1111/1521-9488.00214.

In 2015, Paul F. Diehl and Daniel Druckman also described peacekeeping research as "a-theoretical" and "practically focused."[16] This observation is still largely accurate in 2024 since although the literature on peace operations has grown considerably since the 1990s, only two academic journals are dedicated to peacekeeping (*International Peacekeeping* and the more legal *Journal of International Peacekeeping*), and in international relations journals and works, the preferred approach to dealing with peace operations continues to be largely empirical, aiming to evaluate this or that operation and to demonstrate the performance and effectiveness (or lack thereof) of peacekeeping more broadly.

This observation also applies to the production of *think tanks* on peacekeeping, notably through research projects such as *Providing for Peacekeeping* (co-managed by the *International Peace Institute* in New York, and the *George Washington University* and *University of Queensland*) which ended in 2018,[17] the *Global Peace Operations Review* (conducted by the *Center on International Cooperation* at *New York University*),[18] the program of the *Stimson Center* in Washington DC,[19] or the more recent Effective Peace Operations Network (EPON) managed by the *Norwegian Institute of International Affairs (NUPI)* with 40 other partners around the world.[20]

Other *think tanks* working on peace operations—including the *Folke Bernadotte Academy* (FBA) and the Swedish *Stockholm International Peace Research Institute* (SIPRI), the German *Berliner Zentrum für Internationale Friedenseinsätze* (ZIF), and the *Danish Institute for International Studies* (DIIS) Danish—are financed largely or entirely by the governments of their countries and inform politicians while serving as a tool for influence. Since 2016, the French Ministry of the Armed Forces also co-finances together with the International Organization of *La Francophonie* and the Canadian and Belgian Foreign Ministries) the francophone *Boutros-Ghali Peacekeeping Observatory*.[21] Before that, the *Research Network on Peace Operations* (ROP) created in 2005 and affili-

[16] Paul F. Diehl and Daniel Druckman, *Evaluating Peace Operations*, eds. by Joachim A. Koops et al. (Oxford University Press, 2015), https://doi.org/10.1093/oxfordhb/9780199686049.013.3. P 94.

[17] http://www.providingforpeacekeeping.org

[18] https://peaceoperationsreview.org

[19] https://www.stimson.org/program/peace-operations/

[20] https://effectivepeaceops.net

[21] https://www.observatoire-boutros-ghali.org/node/88

ated with the University of Montreal already had the ambition to promote francophone research on peacekeeping.[22]

The numerous empirical studies produced by these *think tanks* have also been cited by the UN Secretary-General himself to justify the peacekeeping budget and defend its operations against repeated attacks from the US administration of president Trump notably:

> Independent studies have shown the value of peacekeeping operations: they prevent violence from spreading and reduce the number of civilians killed by 90% compared to the period before their deployment. We also know that peacekeeping is effective relative to its cost. The UN peacekeeping budget is less than 0.5% of global military spending, and it is distributed among all 193 member states. In the United States, studies show that peacekeeping missions are approximately 8 times more cost-effective than operations conducted unilaterally by the United States. When you factor in the economic growth and prosperity that comes from the stability and security that results from successful mission deployment, this investment always pays off.[23]

These studies, which Roland Paris describes as "micro" in that they focus on the design, conduct and results of these operations, can provide a better understanding of the circumstances which make an operation successful or not, and make recommendations on how to adapt them. They fuel debates between Member States and within the UN Secretariat. The Division of Policy, Evaluation and Training (DPET) of the Department of Peace Operations and Department of Operational Support (DPO/DOS) effectively serves as a link between the United Nations and the community of academics and *think tanks*. It develops and disseminates the policies and doctrines that govern peace operations and evaluates the way in which these policies are applied so as to complete the learning cycle.[24]

However, only a "macro" and more theoretical approach makes it possible to better analyze and understand these peace operations as the product of the international system, as an instrument and object of power and

[22] The *Peace Operations Research Network* (ROP) has been directed since 2016 by Marie-Joëlle Zahar.

[23] António Guterres, "Peacekeeping is profitable, but must adapt to the new reality," May 2017, https://www.un.org/sg/fr/content/sg/articles/2017-05-30/peacekeeping-cost-effective-must-adapt-new-reality.

[24] "Who We Are - Peacekeeping Resource Center - Research Guides at United Nations Dag Hammarskjöld Library." United Nations. United Nations. Accessed February 12, 2022. https://research.un.org/en/peacekeeping-community/about.

influence. This book therefore aims to focus on the "macro" and to shed light on the relationship between UN peace operations and the theory of international relations.

In 2020, Kseniya Oksamytna and John Karlsrud co-edited a collective volume that examines different issues related to UN peacekeeping by applying different international relations theoretical frameworks (sometimes in a slightly forced manner), while recognizing that "most analyses of peacekeeping, as well as of other issues in International Relations, often draw on several theoretical traditions, rather than one theory." They note that:

> The situation has begun to change recently as a result of two trends: first, peacekeeping scholars have emphasized the connection between their work and the International Relations (IR) literature; second, interest in peacekeeping as a topic for developing and testing IR theories has grown among political scientists. As a result, the main concepts of IR - power, sovereignty, collective action, delegation and gender - have found new applications in the peacekeeping literature, while peacekeeping has become a source of conceptual development and of empirical innovation in the IR literature. This development is late, given the political and material resources invested by member states, international organizations and civil society actors in maintaining peace.[25]

This book, while being primarily an empirical monograph of MINUSMA, also aims to be a contribution to this movement of rapprochement and reconciliation of UN peace operations with political science and international relations (IR). The next three sections of this introduction provide a critical analysis of the dominant liberal approach in the study of peacekeeping since the end of the Cold War, show how the "stabilization turn" since the 2000s has challenged this liberal paradigm, and propose a new "realist-constructivist" approach, to help understand the complex and differentiated modes of action and impact of power rivalries on UN peace operations, both in New York and in the field.

[25] Kseniya Oksamytna and John Karlsrud, *United Nations Peace Operations and International Relations Theory*. (Place of publication not identified: MANCHESTER UNIV Press, 2020). Introduction.

The Dominant Liberal Approach in the Study of Peacekeeping

The liberal approach to international relations largely dominates the literature on the UN and UN peacekeeping. Conceptually first, the UN as an international organization aims to regulate interstate relations by promoting norms and cooperation between states. This growing interdependence aims to reduce uncertainty and thus to enable the pacification of interstate relations and the development of collective security.

The Charter of the United Nations and international law prohibit the use of force in relations between States (Article 2§4),[26] and limit its use to the right of self-defense. In this, the UN materializes the idealist aspirations of Emmanuel Kant in the eighteenth century, of removing States from their "state of nature" that is a state of conflict, through an alliance between sovereign States that collaborate in peace because they are interdependent.

Beyond the forum for cooperation between States that the UN represents, peacekeeping operations represent a tool for the collective use of armed force against those who threaten the peace, that is, a tool for conflict management. It is also, in a way, a "decentralization" of collective security under interim measures under Article 40 of the United Nations Charter (Chapter VII).[27]

Although initially conceived as simple missions to observe a ceasefire or a peace agreement between two or more States, deployed within the framework of Chapter VI of the Charter of the United Nations "peaceful settlement of disputes", the ambition of peacekeeping operations increased at the end of the Cold War with more and more so-called "multidimensional" operations. Since 2004, some of these UN missions have been called "stabilization" operations and deployed within the framework of Chapter VII ("action with respect to threats to the peace, breaches of the peace and acts of aggression"). The UN has also authorized non-UN operations under Chapter VIII ("regional arrangements") led by regional and sub-regional organizations, the most advanced model of UN

[26] Alexandra Novosseloff, The *United Nations Security Council and the control of armed force: dialectic of politics and the military in matters of international peace and security*, International Organization and International Relations 56 (Brussels: E. Bruylant, 2003).

[27] These measures will initially allow the Security Council to authorize observation missions or groups as a subsidiary organ under Article 29 or as a means of investigation under Article 34 of the Charter.

peacekeeping partnerships having been developed with the African Union (AU) in Somalia and Darfur notably.[28]

Multidimensional peace operations deployed will be given the objective not only of maintaining the peace derived from a ceasefire and/or a political/peace agreement between belligerents, but in some cases have been asked to carry out coercive actions under Chapter VII (use of force), and often to help consolidate peace by promoting a liberal peacebuilding approach. Such peacebuilding activities include the organization and supervision of electoral processes, the establishment and/or reform of state institutions, particularly security and rule of law institutions (which can take decades), protection of civilians mandates since 1999, etc. with the support of increasingly large and specialized civilian components within these UN multidimensional operations.

The so-called executive mandates conferred on the United Nations missions acting as interim administrations in East Timor (UNTAET) and Kosovo (UNMIK) at the end of the 1990s thus represent the paroxysm of the UN liberal enterprise. As Jean-Marie Guéhenno explains, "the UN had never been faced with such an arduous task: for the first time in its history, it was asked to effectively take charge of all the responsibilities of a sovereign government in not one, but two places."[29] But this type of temporary trusteeship mandate has not been considered again after. The *Brahimi report* of 2000 had also denounced these overly ambitious mandates which make the UN a civil organization managing post-crises without having the means, while non-UN intervention forces ensure security. Many Member States and members of the Secretariat also saw in these unprecedented governance responsibilities an uncomfortable parallel to colonial experiences.[30]

This expansion—both quantitative and liberal—of UN peacekeeping is part of the developments taking place in the decade 1990–2000, which Bertrand Badie describes as:

[28] Paul D. Williams and Arthur Boutellis, "Partnership Peacekeeping: Challenges and Opportunities in the United Nations-African Union Relationship." African Affairs. 2014.

[29] Jean-Marie Guéhenno, *The fog of peace: a memoir of International peacekeeping in the 21st century* (Washington, DC: Brookings Institution Press, 2015). P290.

[30] William J. Durch, Victoria K. Holt, Caroline R. Earle and Moira K. Shanahan, The Brahimi Report and the Future of UN Peace Operations (Washington DC: Henry L. Stimson Center, 2003).

> that of trivialized interference and that of the defense of the common goods of humanity. The fact of interfering became a right and even a duty: to re-establish peace, of course, which was in conformity with the Charter of the United Nations, but also to provide relief in the name of humanitarian assistance, to also re-establish a failing internal security, to put an end to civil wars, but also to protect human rights, restore democracy, validate elections (or organize them), counter "ethnic cleansing", stop a genocide.[31]

This liberal approach, defended primarily by Western states, certain non-governmental organizations (NGOs), research centers and think *tanks* and to a certain extent the UN Secretariat itself (the UN bureaucracy), suggests that collective security and peaceful relations between states are best served by the promotion of political systems, models of governance and democratic societies.

Thus, the actions of the Security Council, and in particular the creation of peace operations in response to threats to international peace and security, must not be limited to tensions or conflicts between States, but can also result from tensions and conflicts within a State itself. Following this logic, actors in society are now considered fundamental agents of international politics. For Andrew Moravcsik for example:

> The state is not an actor but a representative institution constantly subject to capture and recovery, construction and reconstruction by coalitions of social actors. Institutions and representative practices constitute the crucial 'transmission belt' through which the preferences and social power of individuals and groups are translated into state policy [...] Government policy is therefore constrained by identities, "the underlying interests and power of individuals and groups (inside and outside the state apparatus) that constantly constrain central decision-makers to pursue policies consistent with their preferences."[32]

Much of the criticism of this liberal approach to "democratic peace" is itself part of a liberal, but complementary, approach to peacekeeping. While recognizing that this approach "convincingly demonstrates that a Western and liberal world culture contributes [...] to giving their unique

[31] Bertrand Badie *The great theoretical debates of the decade*, in Hubert Védrine and Pascal Boniface, *1991–2001: ten years that shook the world* (Paris: IRIS: PUF, 2001).

[32] Andrew Moravcsik, "Taking Preferences Seriously: A Liberal Theory of International Politics," *International Organization* 51, no 4 (1997): 513–53, https://doi.org/10.1162/002081897550447. p. 516.

forms to peace interventions", Séverine Autesserre criticizes, for example, the fact that literature devoted to peace liberal only provides a fragmentary vision:

> It often neglects the concrete practices and daily life of international action, the social and epistemological tensions between international actors and the impact, on missions, of public opinion and domestic political considerations. It also regularly ignores that in-depth knowledge of specific case studies represents a necessary condition for any theoretical development. Finally, this literature pays only fleeting attention to the way in which international interventions operate on the ground, that is to say in the rural worlds and provinces where most peacekeepers are deployed.[33]

As part of her study of the United Nations operation in Congo, Séverine Autesserre first notes that "top-down" conflict resolution approaches which rely mainly on national elites do not translate at the local level. Indeed, the instructions issued from capitals and the headquarters of international organizations do not automatically translate into adoption of consistent measures on the ground; local dynamics of war and peace generally differ from that observable at the state level. She thus pleads, in her book *Peaceland*, for a much more in-depth analysis of field dynamics and for a "bottom-up" approach of an anthropological type in order to make peace operations more effective, particularly in their capacity to protect civilians.[34]

Séverine Autesserre suggests in a more recent article that the UN should recognize that resolving local disputes is just as important as dealing with broader issues, and that the Security Council should require all missions to support the consolidation of the peace from bottom-up rather than just top-down deals struck with elites and fixates on elections.[35] It thus echoes the work of Virginia Page Fortna which emphasizes the impact

[33] Séverine Autesserre, "Building peace: collective conceptions of its establishment, its maintenance and its consolidation", *Critique international* n° 51, no [2] (2011): 153, https://doi.org/10.3917/crii.051.0153.

[34] Séverine Autesserre, *Peaceland: Conflict Resolution and the Everyday Politics of International Intervention* (Cambridge, UK: Cambridge University Press, 2014).

[35] Autesserre, Séverine. 2019. "The Crisis of Peacekeeping: Why the UN Can't End Wars." Foreign Affairs, January/February.

of operations on local populations (the 'peacekept')[36] and the numerous calls for more "people-centered" peace operations.[37]

Within this critical liberal approach to peacekeeping, a large number of researchers have long highlighted the gap between international conceptions of conflict management inspired by the liberal peace agenda and the realities and conceptions of peace of the actors and local societies. This is for example the case of Béatrice Pouligny, who already used the examples of Haiti, El Salvador, Bosnia, Cambodia, Somalia and Congo to illustrate the gap between the complex realities of local populations and inexperienced mission personnel.[38] Some authors such as Roger Mac Ginty attempt to synthesize "bottom-up" and "top-down" approaches by conceptualizing a "hybrid" or "composite" peace resulting from the interaction between actors, networks and liberal peace structures and local actors who resist, ignore or adapt to these liberal peace interventions, and maintain alternative forms of peacemaking.[39]

So-called bottom-up or hybrid approaches join an important more sociological or "culturalist" literature. This literature gives an important place to culture in the design and conduct of UN peace operations. Some researchers thus analyze the impact of the culture of different contingents of peace[40] but others such as Roland Paris conclude more broadly that the international normative environment, that is, a "world culture" dominated by the Western liberal approach, shapes peacekeeping in a fundamental way. He argues in particular that even if the perceived legitimacy of different types of states evolves over time, electoral or liberal democracy established itself as the only model of legitimate government at the end of

[36] Virginia Page Fortna, *Does Peacekeeping Work? Shaping Belligerents' Choices after Civil War* (Princeton: Princeton Univ. Press, 2008).

[37] Cedric de Coning, "Towards More People-Centric Peace Operations: From 'Extension of State Authority' to 'Strengthening Inclusive State-Society Relations'," *Stability: International Journal of Security & Development* 4, no 1 (October 29, 2015): 49, https://doi.org/10.5334/sta.gl. See also HIPPO Report.

[38] Béatrice Pouligny, *They promised us peace: UN operations and local populations*, Mondes (Paris: Presses de la Fondation nationale des sciences politiques, 2004).

[39] Roger Mac Ginty, "Hybrid Peace: The Interaction Between Top-Down and Bottom-Up Peace," *Security Dialogue* 41, no 4 (August 2010): 391–412, https://doi.org/10.1177/0967010610374312.

[40] Chiara Ruffa, *Military cultures in peace and stability operations: Afghanistan and Lebanon*, 1st edition (Philadelphia: University of Pennsylvania Press, 2018).

the Cold War, and was thus the model promoted by the UN through its operations.[41]

Others go even further and do not hesitate to characterize peacekeeping as an imperial[42] or neocolonial enterprise, in that it involves foreign forces promoting a set of values aimed at transforming societies.[43] Michael Doyle and Nicholas Sambanis thus suggest that multidimensional peacekeeping involves a "quasi-colonial presence."[44] While the interests of Western colonial powers undoubtedly shaped UN interventions in Suez and the Congo in the 1960s, it is certainly more difficult for former colonial powers to impose their will on other members of the Security Council, troop contributing countries and UN employees today.

However, this does not prevent these same powers from trying to promote their interests notably in Africa, particularly through the creation of new peace operations. France as an important penholder in the Security Council has sometimes been accused of that in its former colonies such as Chad, Mali and the Central African Republic. Roland Paris contests this neocolonial characterization, arguing that unlike European colonialism which was practiced mainly for the benefit of the imperial states themselves, peace operations are less mercenary, are not based on theories of racial superiority and are limited in time. We also observe that the control exercised by a single power over the conduct of an operation is always relative given the multiple and contradictory interests, and the complex games of power, influence and resistance, between member states and within the UN bureaucracy.

[41] Roland Paris, "Peacekeeping and the Constraints of Global Culture", *European Journal of International Relations* 9, no[3] (September 2003): 441–73, https://doi.org/10.1177/13540661030093004.

[42] Michael Ignatieff, *Empire Lite: Nation-Building in Bosnia, Kosovo and Afghanistan*, A Vintage Original (London: Vintage, 2003).

[43] See the analysis of these debates by Philip Cunliffe, "Still the Specter at the Feast: Comparisons between Peacekeeping and Imperialism in Peacekeeping Studies Today," *International Peacekeeping* 19, no · 4 (August 2012): 426–42, https://doi.org/10.1080/13533312.2012.709751.

[44] Michael W. Doyle and Nicholas Sambanis, *Making war and building peace: United Nations peace operations*, Princeton paperbacks (Princeton, NJ: Princeton University Press, 2006). P318.

The "Stabilization Turn" Challenges the Liberal Approach

The previous section reviewed the main critiques of the dominant liberal approach ("bottom-up," "hybrid," and culturalist) which themselves remain part of the liberal paradigm. This section now analyses how the emergence of so-called UN "stabilization" operations after September 11, 2001 challenges the dominant liberal approach in the analysis of UN peacekeeping. Roger Mac Ginty describes stabilization as:

> an essentially conservative doctrine that runs counter to its stated goals of strengthening local participation and legitimacy. This is a control program which favors notions of assimilation to international (Western) norms and integrates the military into peacekeeping operations. As a result, the value of peace is undermined.[45]

John Karlsrud also expresses concern about the loss of momentum of the concept of liberal peace both at the conceptual level and in practice, due to the renewed interest of Western states in UN stabilization operations following long engagements in Afghanistan and in a context of financial crisis. Adam Day and Charles Hunt also denounce a

> concept of stabilisation [that] relies upon a series of binaries — formal/informal actors, licit/illicit activities, governed/ungoverned space — which often distort the far more complex power relations in conflict settings. As a result, UN peace operations tend to direct resources towards state institutions and ignore a wide range of non-state entities, many of which are crucial sources of governance and exist at the local and national level.[46]

The Security Council has already been using the term stabilization for a long time[47] and has mandated four so-called stabilization UN operations since 2004 (MINUSTAH in Haiti, MONUSCO in the DRC, MINUSMA

[45] Roger Mac Ginty, "Against Stabilization," *Stability: International Journal of Security and Development* 1, no 1 (November 1, 2012): 20, https://doi.org/10.5334/sta.ab.

[46] Adam C. Day and Charles T. Hunt, "UN Stabilization Operations and the Problem of Non-Linear Change: A Relational Approach to Intervening in Governance Ecosystems," *Stability: International Journal of Security and Development* 9, no 1 (2020): 2, https://doi.org/10.5334/sta.727.

[47] Curran, D. and Holtom, P, "Resonating, Rejecting, Reinterpreting: Mapping the Stabilization Discourse in the United Nations Security Council, 2000–14," *Stability: International Journal of Security & Development* 4, no 1 (October 29, 2015): 50, https://doi.org/10.5334/sta.gm.

in Mali and MINUSCA in the Central African Republic) while authorizing support for other non-UN stabilization operations such as AMISOM. However, the Security Council has never really defined the term.

One of the reasons is that the term "stabilization" has long been a bit taboo in the UN context, due to its military (before becoming "pan-governmental") and Western origin and its association with the doctrine of counter-insurgency, and the fact that the notion indeed potentially calls into question the fundamental principles of UN peacekeeping.

The 2015 High-Level Independent Panel on Peace Operations (HIPPO) report noted divisions among Member States on this issue: "Some Member States, including several major troop contributing countries, told the Panel that they were firmly convinced of the need to adhere to the three fundamental principles of peacekeeping, namely the consent of the parties, impartiality and non-use of force, except in self-defense or defense of the mandate. Others, however, felt that these principles were outdated and needed to be revised." The HIPPO ultimately concluded inconclusively that, "this term [stabilization] can be interpreted in several ways" and suggested the Security Council give clarification to how the UN interprets 'stabilization.'[48]

Without this clarification having ever been made, the term 'stabilization' has nevertheless gradually established itself in the UN ecosystem, while continuing to be the subject of debate within a very divided multilateral organization, as well as in the more academic literature.

The vast majority of researchers see in these stabilization operations the return to more Westphalian approaches centered on host states and the "restoration" and/or "extension" of their authority (to use the terms used in Security Council resolutions mandating these operations). Many also see a return to more military-centric approaches to UN peacekeeping and more robust military postures with, where appropriate, a possible recourse to the use of force, particularly against non-state actors defying the authority of the host State. This suggests that these "stabilization" operations would be a new type of peacekeeping operations that would challenge the dominant liberal approach of the 1990s. Robert Muggah writes:

[48] Report of the High-Level Independent Panel on United Nations Peace Operations (HIPPO), A/70/95–S/2015/446, paragraph 134.

> In practice, stabilization serves as a synonym for a new kind of peacekeeping mission [...] the United Nations' gradual adoption of stabilization practices appears to be driven as much by internal institutional pressures—including the pursuit of integrated missions—as by a wider appreciation of external threats including fragility, crime, and extremism. United Nations missions are increasingly multidimensional combining robust operations with peacebuilding and statebuilding activities.[49]

Cedric de Coning, for his part, notes that "the essential difference between peacekeeping and stabilization seems to be that the objective of peacekeeping is to achieve and maintain a ceasefire and/or to support the implementation of a peace agreement between parties to a conflict, while in stabilization, the theory of change involves achieving peace by managing or removing an aggressor." He describes "UN stabilization missions as operations aimed at assisting states in crisis to restore order and stability in the absence of a peace agreement, using force and other means to assist national and local authorities in containing aggressors (as identified in relevant Security Council resolutions) and protecting civilians, as part of a broader peace process that seeks a lasting political solution to the crisis."[50]

Similarly, Aditi Gorur observes that the "mandates of the four stabilization missions and the conditions under which they were authorized reveal a common tendency to assist the host state government in reconquering territory controlled by armed non-state actors who have been deemed politically illegitimate by UN member states, the host state government, and/or parties to a peace agreement."[51]

A number of scholars also suggest that the use of force in a more coercive and proactive manner, therefore beyond what the *Capstone doctrine* of peacekeeping envisages (to "defend their mandate, particularly in situations where the 'State is unable to maintain security and public order'"),

[49] Muggah, Robert, Daniel Forti and Mark J. Wood, Fergus Green, Daniel Forti and Priya Swyden, Katherine Chandler, Christian Bueger, and Daniel Forti and Priyal Singh. "The United Nations Turns to Stabilization." IPI Global Observatory, December 12, 2014. https://theglobalobservatory.org/2014/12/united-nations-peacekeeping-peacebuilding-stabilization/.

[50] Cedric De Coning, Chiyuki Aoi, and John Karlsrud, eds., *UN peacekeeping doctrine in a new era: adapting to stabilization, protection and new threats* (London, United Kingdom of Great Britain and Northern Ireland, USA of America, 2017).

[51] Aditi Gorur, *Defining the Boundaries of UN Stabilization Missions* (Washington, DC: Stimson Center, 2016) p. 5.

is what truly distinguishes stabilization operations from other peacekeeping operations.

De Coning suggests that, "a key element that distinguishes these stabilization missions from other peacekeeping missions is that, in each of these missions, the United Nations Security Council has identified the aggressors to be contained. The aggressors are located outside the context of 'parties to the conflict.'[52]

John Karlsrud also notes that "the mandates for the peacekeeping missions in the CAR (MINUSCA), DRC (MONUSCO) and Mali (MINUSMA) amount to a doctrinal change in UN peacekeeping towards stabilisation and peace-enforcement missions" even though he recognizes "that there is no link between the degree to which a UN peacekeeping operation can use force and whether it is labelled a 'stabilisation' mission or not"[53] since other UN peacekeeping operations that are not called "stabilization" are also authorized to "use all necessary means" to fulfill their mandate of protection of civilians.

While largely subscribing to these analyses, we observe that this turning point in stabilization post September 11, 2001 nevertheless goes well beyond the four peace operations that have "stabilization" in their name. It is in fact not so much the stabilization mandates as such which dictates the conduct of peace operations, but rather the evolution of the international context and of power relations between States. We observe that the use of force is not the monopoly of these four stabilization operations and that other peacekeeping operations which do not have "stabilization" in their name have adopted very similar approaches.

MINUSTAH of course used force as part of anti-gang operations in Haiti in 2006–2007, but the operation in Côte d'Ivoire (UNOCI) also used force against forces loyal to the outgoing president Laurent Gbagbo when he refused to leave power following the elections in 2011. MONUSCO's Force Intervention Brigade (FIB) defeated the rebel group "M23" in eastern DRC but became much more passive militarily in the face of other armed groups thereafter.

[52] Cedric De Coning "Implications of stabilization mandates for the use of force in UN peace operations" in Peter Nadin, ed., *The use of force in UN peacekeeping*, Routledge global institutions series (London; New York: Routledge, Taylor & Francis Group, 2018).

[53] John Karlsrud, "The UN at War: Examining the Consequences of Peace-Enforcement Mandates for the UN Peacekeeping Operations in the CAR, the DRC and Mali," *Third World Quarterly* 36, no. 1 (January 2, 2015): 40–54, https://doi.org/10.1080/01436597.2015.976016.

Conversely in Mali, despite the fact that MINUSMA's mandate was made "more robust" in 2016, this did not lead the mission to use force more. The UN mission has actually been repeatedly criticized for precisely that reason by the host state, which wanted MINUSMA to confront armed groups more actively. Finally, in the Central African Republic, MINUSCA was granted Urgent Temporary Measures (UTMs) allowing peacekeepers to "carry out arrests and detentions" "in areas where national security forces or judicial authorities are not present or are not operational"[54] and the mission has, on multiple occasions, used military force, notably thanks to its quick reaction force.

We therefore observe, contrary to what most of the researchers above seem to suggest, that the mandates of so-called UN stabilization operations are initially not very different from those of other UN peace operations. But it is rather a context and particular moments that make the use of force at the tactical level possible and useful as part of a political strategy or in a reactive manner.

It is however true that armed groups and/or certain parties to a conflict have become targeted by certain peace operations for what they are (and they are designated as "spoilers") rather than for what they did, even if mandates to protect civilians often allow a certain ambiguity. This targeting generally results from the political will (and in some cases pressure) from the host government and/or influential members of the Security Council, and of course the will and capacity of the UN operation on the ground and particular its military and police contingents.

This new "peace enforcement" approach that sometimes include the "neutralization" of armed groups by UN peace operations echoes counterterrorism debates raised by the 2015 HIPPO report:

> owing to their composition and character, are not suited to engage in military counter terrorism operations. They lack the specific equipment, intelligence, logistics, capabilities and specialized military preparation required, among other aspects. Such operations should be undertaken by the host Government or by a capable regional force or an ad hoc coalition authorized by the Security Council.[55]

[54] Security Council, Resolution 2217, S/RES/2217 (2015), April 28, 2015.
[55] HIPPO Report, para 119 & 120.

During the 2016 session of the Special Committee on Peacekeeping Operations (C-34) which analyzed the HIPPO report published a few months earlier, divisions between member states surfaced, mainly around this question of counterterrorism. The United States and the European Union have maintained the position that peace operations "are not currently suited" to conduct military counterterrorism operations (thus suggesting that they may be in the future under certain conditions), while members of the G77 /NAM as well as the Latin American regional group ABUM (Argentina, Brazil, Uruguay, and Mexico) were of the opinion that peace operations "should not" engage in this path. In the end, none of the paragraphs dealing with counterterrorism were retained in the 2016 C-34 report due to the lack of consensus.

Beyond counterterrorism, the HIPPO report recognizes that:

> it is the prerogative of the Security Council to authorize United Nations peacekeeping operations to undertake enforcement tasks, including targeted offensive operations, and that it has done so in the past, as in Somalia in 1993 and in the Democratic Republic of the Congo in 2013 [but insists that] extreme caution should guide the mandating of enforcement tasks to degrade, neutralize or defeat a designated enemy. Such operations should be exceptional, time limited and undertaken with full awareness of the risks and responsibilities for the United Nations mission as a whole [including to become a party to the conflict]. Where a parallel force is engaged in offensive combat operations it is important for United Nations peacekeeping operations to maintain a clear division of labour and distinction of roles.[56]

The use of force by peace operations over the past twenty years, in increasingly complex contexts where peacekeepers no longer maintain peace but rather find themselves trying to manage conflicts, goes well beyond what is contemplated by the 2008 *Capstone Doctrine*. Debates on stabilization and counterterrorism often tend to reduce the question of adapting UN peace operations to complex environments (and new threats) to the sole security and military question and in particular to support for security institutions of the host state (and their elites) themselves predatory and/or corrupt and often part of the problem.

These stabilization approaches have important implications for the impartiality of UN missions, and the use of force, which can otherwise open political spaces in certain contexts. They often result in reducing the

[56] Ibid, para 121–123.

space for more political solutions and the space for political rather than military engagement with armed groups. They also limit the possibility of emphasizing governance and more local or "bottom-up" approaches to use the terminology of Séverine Autesserre or "hybrid" approaches advocated by Roger Mac Ginty, and "non-state forms of governance" often more legitimate in the eyes of the population.[57]

This should not come as a surprise since UN peacekeeping has always been a Westphalian enterprise. The United Nations Charter itself promotes an institutional form of the Westphalian state and the centrality of territorial sovereignty, despite the first words of the preamble to the Charter "We, the peoples of the United Nations." But as Roland Paris points out, the international norms promoted by peace operations are not static and evolve as the world order dictates.[58] Historically, these international norms have not only served the powerful states of the Security Council but have also been largely defended by middle powers involved in peacekeeping as an important collective security tool.

The evolution of the world order since September 11, 2001, the end of the American "unipolar moment" which guaranteed a world order the United States was itself at the origin of, and the advent of a world that some describe as "multipolar" with the return of global Russia, the rise of China, and more assertive regional powers with global ambitions, help explain these recent developments in peacekeeping. The powerful states had for the most part lost interest in peacekeeping—particularly following the UN failures of the mid-1990s—and had left the field open to middle powers which have been the main defenders of liberal democratic norms at the UN. But in a changing global order, these powerful states are now reaffirming their diverging interests and values, including through the multilateral system and UN peace operations.

This book offers a "realist-constructivist" approach to the study of MINUSMA and peace operations more generally, that aims to challenge but also complement the dominant liberal paradigm when it comes to the study of UN peace operations. While the analysis of UN peace operations cannot be reduced to only that of geostrategic rivalries, the book

[57] Day, AC and Hunt, CT, 2020. UN Stabilization Operations and the Problem of Non-Linear Change: A Relational Approach to Intervening in Governance Ecosystems. *Stability: International Journal of Security and Development*, 9(1), p. 2. DOI: https://doi.org/10.5334/sta.727.

[58] Paris, Roland Paris, "Peacekeeping and the Constraints of Global Culture," European Journal of International Relations 9, no. 3 (2003), 448–50.

demonstrates that it is not possible to understand historical developments in UN peace operations, including the stabilization turn of the past twenty years, without factoring the evolution of power rivalries.

A "Realist-Constructivist" Approach to UN Peace Operations

The "realist-constructivist" approach adopted in this book is inspired by Raymond Aron and Stanley Hoffmann on the one hand and Samuel Barkin on the other hand. Where the dominant liberal approach denounces developments of the past two decades and the stabilization turn in UN peace operations as a drift toward approaches that are too state and security-focused, our realist-constructivist approach proposes an analytical framework that helps better understand contemporary peacekeeping trends in relation to evolutions in power rivalries.

For Samuel Barkin, "the realist constructivism would look at the way in which power structures affect patterns of normative change in international relations and, conversely, the way in which a particular set of norms affect power structures."[59] This makes it possible to analyze how power rivalries at the UN also play out in the area of ideas, principles, values and norms (impartiality, sovereignty, gender, human rights, protection of civilians, factoring in climate change and the environment, etc.), as well as the consequences this may have in the implementation of mandates in practice.[60]

This "realist-constructivist" approach has the advantage of going beyond the arbitrary distinction between so-called normative theory and so-called empirical theory of world order. It transcends the paradigmatic approach and the "isms" (realism, liberalism, constructivism) which too often characterize international relations, and instead aim to promote a more multidisciplinary approach to the study of the World organization and its peace operations.

Far from opposing constructivism and realism, we suggest like Samuel Barkin that, "constructivist research is compatible with a realistic

[59] J. Samuel Barkin, *Realist Constructivism Rethinking International Relations Theory* (Cambridge; New York: Cambridge University Press, 2012). P. 166.

[60] John Karlsrud, *Norm change in international relations: linked ecologies in UN peacekeeping operations* (London etc., United Kingdom of Great Britain and Northern Ireland: Routledge, 2016).

worldview."[61] Raymond Aron's theory of international relations is also compatible with a constructivist approach,[62] and like Aron, Stanley Hoffmann, who wrote about international organizations that "when solidarity existed, organizations were useless; when it ceased to manifest itself, they were powerless",[63] offers a realistic theoretical approach to international relations on power relations which also highlights the moral and ethical foundations of politics, arguing that interests and values are inseparable.[64]

Similarly Alexander Wendt emphasizes that ideas, values and norms matter, and that even if these do not necessarily come from States (and often find their origins in civil societies and think *tanks* in the UN context), change in the international system ultimately occurs primarily because of member states. We adopt a broad definition of power and interest when analyzing power games, interests and influence strategies at the UN, which may include elements of *soft power*. Power rivalries between member states at the UN cannot in fact be understood solely in terms of military power or material means.

> It is not a question of saying that ideas are more important than power or interest, or that they are independent of them. Power and interest are no less important and decisive than before. It is a question of saying that they have the effects that they have by virtue of the ideas which create them. Explanations based on power and interest presuppose ideas, and, to this extent, are not competitive with explanations based on ideas.[65]

From a constructivist perspective, Vincent Pouliot explains that the "diplomatic power" of a country also plays an important role, and that it is based on the size of its delegation in the permanent mission in New York,

[61] J. Samuel Barkin, *Realist Constructivism Rethinking International Relations Theory* (Cambridge; New York: Cambridge University Press, 2012).

[62] Olivier Schmitt, ed., *Raymond Aron and international relations*, The new international relations (London; New York, NY: Routledge, 2018).

[63] Stanley Hoffmann, "International Organization and the International System," Volume 24, Issue 3, Summer 1970, pp. 389–413. DOI: https://doi.org/10.1017/S0020818300025972.

[64] Stanley Hoffmann, *World disorders: troubled peace in the post-Cold War era* (Lanham, MD: Rowman & Littlefield, 1998).

[65] Alexander Wendt, *Social Theory of International Politics*, Cambridge University Press, 1999. P. 135.

the know-how of its staff, administrative structure, and experience.[66] Guillaume Devin adds the fact that relations between States in a multilateral context such as that which characterizes negotiations on peace operations have the originality of "combining this very formal, even rigid aspect of the negotiation framework with extremely informal practices" and in fact very personalized.[67] These Security Council negotiations of resolutions or declarations have become more and more complicated in recent years, due to exacerbated power rivalries particularly between the permanent members of the Council.

Non-state actors, *think tanks*, NGOs and other consultants financed largely by European middle powers (with varying degrees of independence and more or less close links with certain states) also play an increasingly important role in these games of influence particularly in the field of ideas, to the point of being sometimes described as the "Third UN."[68]

The realist paradigm, even if it is often used to analyze the Security Council, has rarely (or poorly) been applied to UN peace operations. Philip Cunliffe notes that, "where realism has been applied to the study of peacekeeping, it has been in a limited way."[69] He notably criticizes Norrie MacQueen[70] for using concepts (such as "failed states" and "globalization") which do not fit easily into a neo-realist framework, while recognizing his "important insight that the growth of peacekeeping was driven by the collapse of geopolitical spheres of influence and contested war zones that emerged from the Cold War." He also criticizes Laura Neack[71] for her use of realism as a synonym for "selfishness" as opposed to "altruism" to theorize the participation of middle powers in peacekeeping. Cunliffe finally raises the fact that "there has also been no particular effort in these studies to deconstruct and clearly differentiate between different levels of

[66] Vincent Pouliot, *International Pecking Orders: The Politics and Practice of Multilateral Diplomacy*. p. 131.

[67] Guillaume Devin, "Words of diplomats".

[68] Carayannis and G. Weiss, *The "third UN"*.

[69] Philip Cunliffe, "Realism" in Oksamytna and Karlsrud, *United Nations Peace Operations and International Relations Theory*. P39.

[70] Norrie MacQueen, *Peacekeeping and the international system* (London, United Kingdom of Great Britain and Northern Ireland, United States of America, 2006).

[71] Laura Neack, "UN Peace- Keeping: In the Interest of Community or Self?", *Journal of Peace Research* 32, no 2 (May 1995): 181–96, https://doi.org/10.1177/0022343395032002005.

analysis - such as the unit level (state behavior) in relation to the system level (international effects)."

Lise Howard in her book *Power in Peacekeeping* published in 2019, uses Robert Dahl's conceptualization of power to demonstrate that peacekeepers can be effective and achieve their ends in three ways, through persuasion (mainly verbal), incitement (mainly financial) and coercion (mainly through deterrence, surveillance and arrest). However, she does not analyze how power is exercised or not, and how, through the peace operations she studies, namely in Namibia, Lebanon, or the Central African Republic. It is in fact a study of the power—and impotence—of peacekeepers (and therefore again a "micro" study), rather than of how the power of states is exercised through peace operations at the "macro" level to use the terminology of Roland Paris.[72]

The great realist authors such as John Mearsheimer have often been a little disdainful of peacekeeping, which according to them has no role to play in disputes between great powers. They often view the UN as powerless since it cannot resort to coercion, as seen today in crises involving great powers (and increasingly regional powers) in Ukraine, Syria, Libya, etc. However, Mearsheimer recognizes that institutions like the United Nations, although they have little influence on the behavior of powerful states, are essentially a reflection of the distribution of power in the world.[73]

Realism therefore allows us to make the link between peacekeeping dynamics and competition between powers. As Hans J. Morgenthau writes, "we cannot escape the temptation and liabilities of power politics by an act of will. We must learn to live with them and remain civilized; that is, we must make the most of it. Yet even at our best, we cannot afford to forget that they are still with us."[74]

Our approach is realistic in that it submits to the traditional postulates of this school: (1) the anarchy of the international system in which the interests of each side are opposed, all the more so in these times of questioning of the unipolar world order and multilateralism ; (2) the fact that the main actors in the international system remain nation states, which are of course not the only actors but remain the dominant actors and that the

[72] Lise Morjé Howard, *Power in peacekeeping* (Cambridge, United Kingdom; New York, NY: Cambridge University Press, 2019).

[73] Mearsheimer, "The False Promise of International Institutions." 5–49.

[74] Hans J. Morgenthau, 'The Moral Dilemma of Political Action', in Politics in the Twentieth Century, Vol. I: The Decline of Democratic Politics (Chicago: Chicago University Press, 1971), p. 13.

effectiveness of international institutions such as the Security Council depends on their compliance with interests of the great powers; and (3) states seek to maximize their national interest, including through these institutions; (4) foreign policy generally takes precedence over domestic policy,[75] even if the latter and public opinion in particular can impact it as Hoffmann suggests by emphasizing that democracies cannot act externally against the principles they defend inside, particularly on the question of the defense of human rights. The constructivist dimension is also found here with the fact that ideas can influence the power behavior of States.

We observe, for example, how certain Western public opinions disinterested in external military operations or worried by the COVID-19 crisis can contribute to a nationalist withdrawal, a rise in populism and authoritarianism. Europe has notably been weakened by internal divisions and *Brexit,* and it is struggling to play its role as defender of the liberal international order in general and of some of its norms that are under threat, such as universal human rights.[76]

The search for power on the international stage is an end in itself for States even if they participate fully in international institutions such as the UN. And even if the multilateral system is seen by middle powers[77] as a means of regulating the *hubris* of the great powers and the anarchy of the international system, it does not erase the reality of power. As Raymond Aron writes, realism is an approach that encourages "to keep a cool head, to observe the world as it is, and not to imagine it as we would like it to be, to be wary of abstractions."[78]

The renewed games of power and influence against a backdrop of increasingly divergent interests and values at the UN reflects a changing world order and of course results in increased rivalries and tensions between major powers within an increasingly divided Security Council. These tensions increasingly contaminate debates on peace operations, their mandates and their budgets—debates that were once consensual.

[75] Jean-Baptiste Jeangène Vilmer, *Theories of international relations*, What do I know? (Paris cedex 14: Presses Universitaires de France, 2020), https://www.cairn.info/theories-des-relations-internationales%2D%2D9782130785644.htm. p. 24–25.

[76] Stephen Hopgood, *The endtimes of human rights* (Ithaca: Cornell University Press, 2013).

[77] Roland Paris, "Can Middle Powers Save the Liberal World Order?" (Chatham House, June 2019).

[78] Raymond Aron, "In search of a philosophy of foreign policy", *Revue française de science politique* 3, no 1 (1953): 69–91, https://doi.org/10.3406/rfsp.1953.452691. p. 82.

Our approach is also constructivist in that we observe that the influence and interests of States are also increasingly played out in terms of norms and values, and that the interests and identities of the actors (States and their diplomats, the UN bureaucracy and its employees, but also other peripheral actors in *think tanks* in particular) evolve over time and are shaped by their interactions in New York as in field operations, but also by the interactions between these two levels—UN headquarters and the field.

While recognizing the primacy of interstate relations in our analysis of the renewed games of power and influence in UN peace operations and following our "realist-constructivist" theoretical framework, we also use bureaucratic methods and political psychology to explain in particular how the power games that we analyze have different consequences at the level of UN headquarters in New York, and on peace operations on the ground.

The bureaucratic approach developed by Morton Halperin[79] introduces the importance of internal struggles between individuals and organizations defending competing interests, and identifies five objectives of any organization: to justify its existence, to defend or even extend its territory, to maintain its autonomy and the morale of the troops, and obtain an increase in its budget.[80]

This bureaucratic method, which echoes organizational sociology and systemic constructivist, allows us to better understand how and why the United Nations Secretariat (the "UN bureaucracy") does not necessarily reflect the evolution of the power rivalries between states, sometimes because of resistance or simply bureaucratic inertia or because it makes its own proposals on a subject. Indeed, the UN Secretariat in New York and the field missions are in practice far from being simple implementers of Security Council decisions and have a certain autonomy with regard to the intentions (often themselves various) of Member States.[81] Lise Howard notes that field missions are often more effective when they have greater autonomy from UN headquarters, allowing civilian and military personnel to adapt to realities on the ground.[82]

[79] Graham T. Allison and Morton H. Halperin, "Bureaucratic Politics: A Paradigm and Some Policy Implications," *World Politics* 24, no · S1 (1972): 40–79, https://doi.org/10.2307/2010559.

[80] Jeangène Vilmer, *Theories of international relations*. p. 97.

[81] Michael N. Barnett and Martha Finnemore, "The Politics, Power, and Pathologies of International Organizations," *International Organization* 53, no · 4 (1999): 699–732, https://doi.org/10.1162/002081899551048. pp. 699–732.

[82] Howard, *UN Peacekeeping in Civil Wars.*

The contribution of political psychology (the work of Robert Jervis in particular[83]) also allows us to highlight the importance the personality of certain leaders has on decisions and orientations taken under influence by certain member states. This plays out at the level of the Secretary-General[84] himself but also with special representatives who lead peace operations on the ground, and the Force Commanders in these same operations. David Ambrosetti also suggests that in order to understand what leads certain States to engage in cooperation and/or rivalry in matters of international peace and security at the UN, it is necessary to observe the senior officials responsible for representing these states in multilateral arenas at the UN when immersed in repeated interactions with their counterparts.[85]

Powerful states have also recognized the importance of strategies of influence within the bureaucracy itself, in particular by placing their own nationals in key positions both at headquarters and in the field. Some UN officials in some cases play these power rivalries in their own bureaucratic battles, notably between the Department of Peace Operations (DOP) and the Department of Political Affairs and Peacebuilding (DAPCP), but also to guide or disorient certain organizational reforms initiated. Some UN employees also demonstrate "resistance" either through activism, simple aversion to risk or change, or feel the need to "bend the rules" in order to do their job or advance their careers or simply try to exist. All this can lead to a gap between discourse and action, between headquarters and the field, or between policy and practice, which has sometimes been described as "organized hypocrisy."[86]

Jean-Marie Guéhenno paints a vitriolic portrait of "stifling rules that only seasoned bureaucrats could understand, and a bureaucratic culture that rewarded passivity and punished risk-taking, meant that strong personalities rarely stayed very long at the UN. And those who were not

[83] Robert Jervis, *Perception and misperception in international politics*, new edition (Princeton, New Jersey: Princeton University Press, 2017).

[84] Benjamin Rivlin and Leon Gordenker, eds., *The Challenging role of the UN Secretary-General: making "the most impossible job in the world" possible* (Westport, Conn: Praeger, 1993).

[85] Chalmers Larose, "David Ambrosetti, 2009, Diplomatic norms and rivalries at the UN. The Security Council in Hearing, coll. Perspectives on the international, no. 8, Brussels, PIE–Peter Lang, 352 p.", *International Studies* 41, no^2 (2010): 290, https://doi.org/10.7202/044630ar.

[86] Diana Felix da Costa and John Karlsrud, "'Bending the Rules': The Space between HQ Policy and Local Action in UN Civilian Peacekeeping," *Journal of International Peacekeeping* 17, no 3–4 (2013): 293–312, https://doi.org/10.1163/18754112-1704007.

discouraged and stayed often completely ignored the rules, which was also dangerous." At the same time, however, he recognizes the importance of figures like Ralph Bunche, Brian Urquhart, Sergio de Mello, Lakhdar Brahimi, "individuals who make the most of imperfect arrangements. They are more important in the United Nations than in a national bureaucracy, because they are the ones who carry the flame." He adds that the UN Secretariat "actually has much more power than is generally assumed, and with the exception of a few crises where a major power has a strategic interest, the Secretariat can often be in charge as a sixth permanent member of the Council."[87]

It is therefore a mixed "realist-constructivist" approach also using bureaucratic methods and political psychology, which we propose throughout this to tell the story of MINUSMA and how power rivalries in New York and in Mali have played out through different phases in the history of the mission in a context of hardening of international relations. UN peace operations in general, and MINUSMA in particular, have become a new laboratory of rivalries between powers, with different effects. Peace operations are the product of the international system, both instruments and objects of power, sometimes multipliers and sometimes regulators of power, with bureaucratic and political psychology elements further complicating these complex relationships between different member states, the UN Secretariat and the "third UN."

Power and UN Peace Operations: Constraint, Influence and Resistance

The notion of power has accompanied the evolution of the analysis of international relations and continues to be a central notion, and we use in this book the definition by Raymond Aron which is "the capacity of a political unity to impose its will on other units." According to Aron, power is both the goal and the means of foreign policy and is defined as a "potential for influence, command and constraint."[88] Power criteria have largely evolved, since the military and economic resources of a state are no longer sufficient material criteria to explain the relations of power between States,

[87] Guéhenno, *The fog of peace*. p. 307 and p. 312.

[88] Raymond Aron, *Peace and War among Nations with a presentation by the author* (Paris: Calmann-Lévy, 2004). 1962 p. 58.

especially when military power has shown and continues to show its limits from the Middle East to Afghanistan and the Sahel.

If the criteria of coercion, commonly called "hard power", remains relevant, it is now also necessary to take into account the notion of "soft power" as theorized by Joseph Nye to explain the power of attraction of the American cultural model. Soft power falls into three broad categories: culture, political values, and public policies. Authoritarian regimes also make extensive use of *"soft power,"* which is then confused with "*sharp power.*"[89] The particularity of "*sharp power*" is that it characterizes the ability to manipulate public opinion, in particular through the spread of *fake news* and misinformation. It is no longer a question of simply promoting a model but also of disrupting that of the adversary.

The redefinition of the notion of power in recent decades has been accompanied by the emergence of modes of action and strategies close to what is today qualified as influence policy or strategy. Influence aims for a State or a group of States to induce other actors to act in accordance with its interests, principles and values, either by establishing an international order (or norms) which frame or guide their actions, or by establishing the conditions for an objective community of views or interests, that is to say, to make them change their behavior without having to resort to coercion.[90] Influence differs from "*soft power*" in that actions of influence may be more ad hoc and may include compensation or reward.[91]

The ability to influence international institutions such as the UN thus plays a growing role in global power and no longer depends solely on the "*hard power*" of a state but also on its rank and status.[92] Global power is exercised today less through coercion than through influence, that is to say the ability to influence the behavior of other States toward its interests. As Anne Clunan explains, "the status of a state is fundamentally different from its material position in the international system, because status is a product of social construction." It can be defined as "the collective beliefs concerning the ranking of a given State on valued attributes (wealth,

[89] Christopher Walker and Jessica Ludwig, "The Meaning of Sharp Power: How Authoritarian States Project Influence," Foreign Affairs, 16 November 2017.

[90] Bastien Nivet, "Power or influence? A detour through the European experience", International and Strategic Review n°89, Paris, IRIS-Armand Colin, spring 2013, pp. 83–92.

[91] Frédéric Charillon, *Wars of Influence: States Conquering Spirits* (Odile Jacob, 2022).

[92] Serge Sur, "Power and Rank Revisited" in AFRI, Volume I, 2000 06/17/2022 00:18:00.

coercive capacities, culture, demographic position, socio-political organization, diplomatic weight). Status is collective, subjective and relative.[93]

On this basis, we propose our own framework for analyzing the notion of power and its modes of action in the ecosystem of UN peace operations, which we break down into:

- Constraint (any element of pressure, including economic and military),
- Influence (convincing and mobilizing, including through alliances), and
- Resistance (to thwart the will of the powerful, including through sabotage or spoiling).

Influence is particularly important in the UN context, since it involves the use of indirect and non-coercive means and network strategies, whereby the actor around whom the greatest number of strong interdependencies (economic, political, security, etc.) has a magnified influence. Some networks support "*soft power*" of cultural or scientific influence, language—the *Francophonie* network for France, even if this is less and less obvious in a context of rising anti-French sentiment in French-speaking Africa.[94] These *"soft power"* capacities do not replace that of "***hard power***" but complement them, and their articulation is what constitutes the power and influence of States today.

Although it is not always easy to distinguish between power and influence, and the two are linked, in the multilateral system, influence is also often the weapon of "middle powers" or "declining great powers" which no longer necessarily have the means to sustain a strategy based on constraining others, the effectiveness of which is in any case called into question. Maurice Vaïsse thus suggests that General de Gaulle's style of "grandeur" of France was a compensation for the country's strongly relativized power.[95] Influence cannot be seen as the only alternative to constraint (and "so*ft power*" does not always translate into political influence)

[93] Anne L. Clunan, "Why Status Matters in World Politics", in *Status in World Politics*, ed. by TV Paul, Deborah Welch Larson, and William C. Wohlforth (Cambridge: Cambridge University Press, 2014), 273–96, https://doi.org/10.1017/CBO9781107444409.017.

[94] Fanny Pigeaud. "Africa and anti-French sentiment", Bertrand Badie ed., France, a thwarted power. The State of the World 2022. Discovery, 2021, pp. 189–195.

[95] Maurice Vaïsse, *Power or Influence ? France in the world since 1958* (Paris: Fayard, 2009).

but must rather be considered as an alternative or complementary mode of action in an interdependent world.

Multilateralism implies a certain institutionalization of the cooperation between actors.[96] This cooperation produces "norms and rules aimed at establishing a cooperative international order governing international interdependencies"[97] without of course evading the political dimension of these norms which is precisely at the heart of the current crisis of multilateralism. The institutionalization of certain norms over time does not therefore prevent them from being called into question, in particular if the context and the balance of power between blocks of States with given identities and interests evolve.[98]

As Vincent Pouliot suggests however, "the multiple inequalities that structure world politics are not the simple reflection of the distribution of material power from one country to another"[99] and of the balance of power as would tend as suggested by some realist theorists like John Mearsheimer.[100] Pouliot explains that "diplomatic power" also plays an important role, and that it is based on the size of a delegation in the permanent mission in New York, and the know-how and experience of its staff in both formal and informal settings.[101] And even the representatives of great powers like the United States and China wield charm and intimidation. Admittedly, states seek to maximize their relative power, but the

[96] Bertrand Badie and Guillaume Devin (eds.), *Multilateralism. New forms of international action*, Paris, La Découverte, 2007.

[97] Franck Petiteville, *Multilateralism*, Paris, Montchrestien, 2009.

[98] The norm is generally defined as a standard of appropriate behaviors accepted by actors with the given identity. Martha Finnemore and Kathryn Sikking explain the nature, emergence, evolution and influence of norms in international relations, functions of states' need for legitimacy, the quality of states that support a norm, the intrinsic quality of this standard, the path traced by similar standards and ultimately the global context. In this, peace operations can themselves be seen as an accepted norm that has also evolved over the past seventy years. See Martha Finnemore and Kathryn Sikking, "International norm dynamics and political change", International Organization, 52 (4), pp. 887–917.

[99] Vincent Pouliot, "Conclusion. Miracle and Tragedy of International Pecking Orders: The Politics and Practice of Multilateral Diplomacy", *International Pecking Orders: The Politics and Practice of Multilateral Diplomacy. Struggles for Rank in Multilateral Diplomacy*, edited by Pouliot Vincent. Sciences Po Press, 2017, p. 163–180.

[100] John J. Mearsheimer, "The False Promise of International Institutions," *International Security* 19, No.[3] (1994): 5, https://doi.org/10.2307/2539078.

[101] Vincent Pouliot, *International Pecking Orders: The Politics and Practice of Multilateral Diplomacy: Rank Struggles in Multilateral Diplomacy*, International Relations Collection 7 (Paris: Presses de Sciences Po, 2017). p. 131.

"processes for constructing the international hierarchical order are complex" and changing. As Mélanie Albaret notes:

> The links between UN multilateralism and power are not unequivocal and their interpretation cannot be reduced to the instrumentalization of the first by the second. The UN can be a brake or a multiplier of power, a resource and a constraint for powers, an instrument of power and an object of power. Powers can be creators of multilateralism or destroyers of multilateral institutions; they can serve multilateralism, make use of it or do it a disservice.[102]

Multilateral diplomacy does not content itself with reflecting or reproducing inequalities between States; it also engenders them and sometimes modifies them.[103] The middle powers and smaller states can, for example, act in resistance against the most powerful through games of alliances in the Security Council and in the General Assembly, but also by posing as a force of proposal. State power and influence strategies can also be constrained by logics of resistance or simply bureaucratic inertia specific to the United Nations Secretariat and its field operations.

Host governments of peace operations can also leverage their consent, and use their state sovereignty to resist pressure from the Security Council and the UN mission on the ground, as they strive for greater oversight of how and where missions operate.[104] This can be termed the "power of the weak" since states that hosts UN peace operations generally do so because of periods of conflict, violence and fragility they go through at a certain moment. The resistance of host states generally however has limits. Robert Keohane explains that:

> Even if weak states have the nominal capacity to thwart the will of the strong in international organizations, they may be unable to analyze complex issues or make their voices heard. They may also be reluctant to challenge their powerful neighbours, creditors and business partners. Consequently, the apparent diversity of interests in a world Organization can only be nominal. Defenders of multilateral institutions must recognize this intrusion of power

[102] Mélanie Albaret, "The UN, between power and multilateralism", *CERISCOPE Power*, 2013.

[103] Vincent Pouliot, *International Pecking Orders: The Politics and Practice of Multilateral Diplomacy: Rank Struggles in Multilateral Diplomacy*, International Relations Collection 7 (Paris: Presses de Sciences Po, 2017). P. 22 & P. 163.

[104] Julie Gregory and Lisa Sharland, "Host-Country Consent in UN Peacekeeping," The Stimson Center, 2023, Washington, D.C., USA.

politics as an imperfection and a potential threat to the legitimacy of their decisions. But this is not a new problem. The creators and defenders of these organizations in the 20th century were not unaware of power politics.[105]

The UN was born from the projects of great powers and in particular the United States, and from the balance of power of 1945 at the end of the Second World War, reflected in a Security Council in which the five permanent members (or P5 - China, the United States, Russia, France and the United Kingdom) possess both the "right of veto" and the three Western countries among them most often hold the pen—and therefore draft and lead the negotiation—for resolutions mandating peace operations. However, the balance of power within the UN has evolved, with decolonization in particular and the fact that the new independent States also seek to widen their margins of political sovereignty through "non-alignment," using rivalries between different sides.

Due to the relative absence of the P5 from peace operations in the field (among the blue helmets) and the questioning of the effectiveness, representativeness and therefore the legitimacy of the Council, but also to developments in multilateralism which has become more inclusive of civil societies notably,[106] great powers see their dominance and the traditional partnerships at the foundations of peace operations increasingly challenged by other less powerful states and certain non-state actors. With that said, the UN, despite the preamble to the Charter which begins with "We the peoples of the United Nations determined to save succeeding generations from the scourge of war,"[107] remains above all an interstate organization in which states ultimately make decisions including on peace operations.

The dominance of the P5 is increasingly challenged by other powerful and less powerful member states, among the elected members of the Security Council but also within the General Assembly. The role and power of groups of influential states in the General Assembly manifests itself more in terms of resistance—and often through more or less formal

[105] Robert O. Keohane, "Global Governance and Legitimacy," *Review of International Political Economy* 18, No.[1] (February 7, 2011): 99–109, https://doi.org/10.1080/09692290.2011.545222.

[106] Antonio Guterres, "To be more effective, Multilateralism must be more inclusive" | UN News. https://news.un.org/en/story/2018/11/1028791.

[107] "The Charter of the United Nations": *La Pensée* N° 377, no[1] (January 1, 2014): 141–56, https://doi.org/10.3917/lp.377.0141.

Actors & Spaces / Modes of action	***United Nations* *Security Council* *(P5 members in particular)***	***United Nations* *General Assembly***	***United Nations* *Secretariat* *(and its field operations)***
Constraint **(any element of pressure, including economic and military)**	Drafting and negotiating **peace operations (PO) resolutions /mandates (holding the "pen"), Presidential statement, Press releases, letters, etc.**		**Failure to respect the independence of Under Secretary Generals** DPO and DPPA **Politicization of appointments of special envoys and special representatives** of the Secretary-General heading PO
Influence **(convincing and mobilizing, including through alliances)**	**Sharing the "pen"** with an E10 ("co - pen") **Field visits by the Security Council to convince** Approving a **Chapter VII mandate for a regional force** (AU, G5, etc.)	Approving **UN logistical and financial support for a regional force** (AU, G5, etc.)	**Sending "gratis personnel"** (JPOs or seconded staff) to headquarters **and /or** within field operations (staff officers in particular)
Resistance **(to thwart the will of the powerful, including through sabotage)**	**Blocking** the renewal of a peace operation's mandate and/or a proposed draft resolution, or simply proposing an alternative text	**Exercising pressure on PO budgets** (or certain items) & headquarters in 5th Committee **Normative (re)framing** in the Special Committee on Peacekeeping Operations (C-34)	**Resistance and/or bureaucratic inertia** at headquarters and within PO in the face of member states power plays and influence **Building ad hoc coalitions of Member States** to support specific topics/interests

Fig. 1.1 Key actors, spaces and modes of action within this UN ecosystem

The "Third UN" (Think Tanks and NGOs)	***Host government***	***Member States including Troop Contributing Countries (TCCs)***
	Non-consent to mandate (or obtaining concessions in PO format) **Or** **SOFA violations or outright withdrawal of consent** (which usually leads to departure from the PO)	**Deployment of an** operational "parallel force" capable of influencing or constraining a PO Threat of and/or **withdrawal of peacekeeping contingents** **Voluntary/extrabudgetary financial contributions** to special funds at UN headquarters **or** directly to trust funds managed by the PO
Funding Think Tanks and NGOs as a roundabout way to advance analysis of situations, as well as certain political and normative agendas **Advice from Think Tanks and NGOs to certain members of the E10** and other Member States on subjects on the agenda		**Promotion of specific issues and topics** through meetings and conferences **Creation of "Group of friends"** on a subject related to PO **Bilateral support** to certain allied TCCs (logistical, financial and/or capacity building/training) **Influence operations** to shape host state consent
	Obstructing the implementation of the mandate and/or the freedom of movement of a PO (non-compliance with the SOFA) **Rejection of a special representative appointment, boycott, and/or request for the departure** of a UN official from the country (PNG)	**Lack of interest in a PO**, minimal contribution and/or preference for contributions to "parallel forces" (training missions for example) **Resistance of TCCs** to sending peacekeepers and/or implementing a mandate (caveats) and following orders

Fig. 1.1 (continued)

coalitions and alliances. This may involve pressure on budgets (or certain posts) of a peace operation in the Fifth Committee from major or minor financial contributors; a normative (re)framing at the Special Committee on Peacekeeping Operations (C-34) in which the G77 /NAM States weigh in particular.

Non-State actors, *think tanks*, NGOs and other consultants—sometimes called the "Third UN"[108]—funded largely by European middle powers (with varying degrees of independence and more or less close ties with certain States) also play an increasingly important role in these games of influence. States may fund particular programs and projects to attempt to advance certain policy and normative agendas. The most effective initiatives often combine this type of funding with Member State engagement in various informal and formal UN forums over several years.

Power is a whole and there are many modes of actions—including constraint or coercive power, influence, and resistance—which different member states and group of member states, but also non-state actors, can use in more or less effective ways in different formal and informal settings related to the conduct of UN peace operations. The table below intends to summarize key actors and modes of action within this UN ecosystem (Fig. 1.1).

[108] Tatiana Carayannis and Thomas G. Weiss, *The "third one"*, 1st ed. (New York: Oxford University Press, 2021).

CHAPTER 2

MINUSMA First Act (2013–2015): "Multilateralizing" a French-African Intervention

"I had strong doubts about the relevance of sending [to Mali] contingents that I knew would have neither the means nor even the will to fulfill the ambitious mandate I had the Security Council adopt. I actually only overcame the objections from my colleagues [the other ambassadors to the UN] by promising the support of French forces if things went wrong."[1]

Gérard Araud, Former French ambassador to the UN, 2019

The Genesis of the UN Mission in the Wake of French-African Intervention

The background to the deployment of the UN mission is that of a Tuareg rebellion by the *Mouvement national pour la libération de l'Azawad* (MNLA) in January 2012, the fourth since the country's independence in 1960. The rebels, bolstered by the return of soldiers from Gaddafi's legions from Libya with weapons, occupied a large part of northern Mali and declared the independence of this territory, which they call *Azawad*. On January 17, 2012, the MNLA and a Tuareg Islamist group called *Ansar Dine* joined forces to attack a Malian army garrison in Aguelhok, a commune in the Kidal region. The news of the execution of around a

[1] Gérard Araud, *Passeport Diplomatique*, Paris, Grasset, 2019, p. 81.

A. Boutellis, *The UN Stabilization Mission in Mali (MINUSMA)*, Twenty-first Century Perspectives on War, Peace, and Human Conflict, https://doi.org/10.1007/978-3-031-66466-3_2

hundred Malian soldiers whose garrison had surrendered after running out of ammunition on January 24, deeply shocked many Malians.[2]

The poor performance of the Malian armed forces in the face of the rebellion led in turn to a coup d'état in March 2012, led by frustrated elements of the Malian army in the capital Bamako. The coup marked the collapse of a Malian state that had been weakened by corruption for years and President Amadou Toumani Touré (ATT)'s policy of "demilitarization" in northern Mali. ATT was also known for his policy of alliances with certain local elites and ethnic militias in northern Mali. The Tuareg occupation of northern Mali was short-lived, however. Al-Qaeda in the Islamic Maghreb (AQIM), militarily and financially superior, and its allies from the Movement for Unity and Jihad in West Africa (MUJAO) quickly ousted the MNLA elements, taking control of Timbuktu and Gao, while *Ansar Dine* took control of Kidal.

Following the occupation of northern Mali by these Islamist forces, the President of Burkina Faso, Blaise Compaoré, appointed by Economic Community of West African States (ECOWAS) as mediator in the Malian crisis in March 2012, opened negotiations between the interim Malian Government and two of the armed groups, *Ansar Dine* and the weakened MNLA. Algeria also tried to get *Ansar Dine to* disassociate itself from AQIM and MUJAO.[3] These attempts at negotiation were not successful and the Islamist groups reportedly began moving south toward the capital Bamako (although this scenario has since been called into question[4]), and France suspected a new coup d'état was being prepared in Bamako against President Dioncounda Traoré.

This is the context in which France launched its *Serval* military intervention on January 11, 2013, following an explicit request from the Malian transitional government interim president Dioncounda Traoré. Security Council Resolution 2085 (2012) which had been adopted a few days earlier, on December 20, 2012, gave this intervention the international legal cover, as it recalls

> the letter that Mali's transitional authorities sent to the Secretary-General on September 18, 2012, requesting that an international military force be

[2] Jeune Afrique, "Guerre au Mali: retour sur le drame d'Aguelhok", October 21, 2013.

[3] France 24, "Ansar Dine and MNLA ready to cease hostilities and negotiate with Bamako", December 21, 2012.

[4] Jean-Dominique Merchet, "Mali Les colonnes jihadistes fonçant sur Bamako en 2013, une légende!" *Secret Défense*, February 12, 2021.

authorized, under the provisions of Chapter VII of the United Nations Charter, to assist the Malian Armed Forces in retaking the occupied regions in the north of the country.[5]

France garnered the broadest possible political support it could at the UN.[6] Because the African-led International Support Mission to Mali (AFISMA) authorized under Resolution 2085 was slow to deploy, it was eventually overtaken by events and by the French military intervention. After killing a few hundred jihadists in the first few days of operations, and some initial confusion as to the objective of the *Serval* intervention (simply to prepare for the deployment of AFISMA versus a more ambitious objective), French forces rapidly reconquered the northern towns of Gao and Timbuktu, then Kidal with the support of Chadian troops (who lost dozens of soldiers in the fighting), with the Malian army following in their footsteps.[7]

Armed groups put up little resistance, disappearing into the rugged terrain of the *Adrar des Ifoghas* hills in northern Mali and southern Libya. At the time, there was still talk of AFISMA deploying up to 10,000 troops. But the African Union (AU) and AFISMA troop-contributing countries realized that without logistical and financial support from the UN (of the kind the Security Council had granted to AMISOM in Somalia but did not wish to replicate in Mali) and donors[8] they would not be able to sustain such an operation over the medium term. This created some tensions between the UN Security Council and the African Union (AU). Although the AU formally expressed its support for the planned transformation of AFISMA into a UN peace operation, as it did not have the financial and logistical means to support such an operation,[9] it later criticized on multiple occasions the blue helmets for not more actively fighting terrorists in Mali and the fact that a parallel French force (rather than an African force) was in charge.

[5] Security Council Resolution 2085, S/RES/2085 (2012), December 20, 2012.

[6] Natalie Nougayrède, "The Malian president's letter, legal basis for French intervention." Le Monde, January 25, 2013.

[7] Crisis Group, "Mali: sécuriser, dialoguer et réformer en profondeur", April 11, 2013.

[8] AFISMA fund-raising conference held in Addis Ababa at the end of January 2013 was a failure, with no new funding beyond that already pledged by the EU and the United States far from sufficient to sustain the operation over the long term.

[9] African Union, "358th meeting of the Peace and Security Council of the African Union (AU) on the situation in Mali.", March 7, 2013.

Around the same time, African countries and the AU were pushing for the creation of a *Force Intervention Brigade* (FIB) in the DRC, with an offensive mandate and seasoned, well-equipped troops (notably from South Africa and Tanzania) capable of proactively neutralizing certain armed groups operating in eastern DRC. While the Security Council did authorize it "on an exceptional basis and without creating a precedent," the FIB was eventually attached to the United Nations mission in the DRC, MONUSCO, as the UN Secretariat and the Security Council considered that it would be preferable to have a single international force rather than a parallel force under a different chain of command.[10]

Another source of tension was the fact that the UN did not accept the AU's request to appoint an African to head the future UN peacekeeping operation.[11] The Security Council and France in particular, preferred the Dutchman Bert Koenders, previously Special Representative of the Secretary-General and head of the United Nations Operation in Côte d'Ivoire.

The reference in Security Council Resolution 2100 of April 25, 2013 to the fact that only AFISMA troops that "comply with United Nations standards" (paragraph 7) will be integrated into MINUSMA was another source of discord, with African troop-contributing countries feeling devalued. The main African troop contributor, Nigeria, decided to withdraw its troops from Mali after its general failed to secure the post of Force Commander, which went to Rwandan general Jean Bosco Kazura.[12] The African elected members of the Security Council at the time were Rwanda, Togo and Morocco.

Finally, the AU also felt sidelined in the political negotiation process, which would now be "facilitated by the Secretary-General [of the UN], notably through his Special Representative for Mali when appointed … in

[10] A parallel African force that would have neutralized the M23 armed group (which had taken the town of Goma a few months earlier despite the presence of 1000 peacekeepers) also risked raising questions about the usefulness of a UN operation perceived as having lasted too long and being too passive. Boutellis, "From Crisis to Reform."

[11] On January 30, the AU Chairman appointed Pierre Buyoya, already AU High Representative for Mali and the Sahel, as his Special Representative and Head of AFISMA. and the Sahel, as his Special Representative and Head of AFISMA, in the hope that he would become the UN's SRSG in Mali and Head of MINUSMA.

[12] Mouhamadou Kane, "After a Rwandese general is appointed to head MINUSMA, Nigeria withdraws its troops. Why?", ISS, November 15, 2013, https://issafrica.org/iss-today/nigeria-pulls-out-of-minusma-protest-or-prioritisation-of-domestic-security.

close collaboration with the AU, ECOWAS and the EU."[13] This was all the more sensitive in a context in which the AU had already felt sidelined in Libya at the time of the NATO military intervention (the "no fly zone" came into effect the day the AU Ad Hoc Committee had planned to arrive in Tripoli).[14] Later in 2019, the request by the three African members (A3) of the Security Council to appoint a joint UN-AU special envoy for Libya was rejected Council's permanent members.[15]

This combination of factors explains why, on the day the Security Council adopted Resolution 2100, despite a statement of support for "the decision to transform AFISMA into a United Nations stabilization mission UN stabilization mission"[16] by the African Union Observer in New York, the African Union Peace and Security Council denounced the opacity and speed with which the UN had taken its decision, noting with concern:

> that Africa was not appropriately consulted in the drafting and consultation process that led to the adoption of the UN Security Council resolution authorizing the deployment of ... MINUSMA to take over AFISMA, and stresses that this situation is not in consonance with the spirit of partnership that the AU and the United Nations have been striving to promote for many years, on the basis of the provisions of Chapter VIII of the UN Charter.[17]

Despite these frictions with the region, France ultimately had little difficulty in convincing the Security Council to unanimously approve the creation of what would become the UN's largest and most expensive peacekeeping operation.

[13] Security Council Resolution 2100, S/RES/2100 (2013), April 25, 2013, para 4.

[14] The AU claimed that Muammar Gaddafi had accepted its "roadmap" for a negotiated way out of the crisis. The AU has never really come to terms with this episode, which continues to pollute UN-AU relations on Libya. Jean Ping: "les pays occidentaux n'ont pas laissé l'Union africaine 'entrer' dans cette histoire libyenne", l'Humanité, June 14, 2014 and Radio France Internationale (RFI), "South African President Zuma castigates the West at the UN on the Libyan dossier," January 13, 2012.

[15] United Nations, "African members of the Security Council call for enhanced cooperation between the Council and the African Union, including on Libya - press releases." CS/14001, October 30, 2019.

[16] Security Council, S/PV.6952, April 25, 2013.

[17] African Union, "Communiqué - The Peace and Security Council of the African Union (AU), at its 371st meeting held on 25 April 2013, on the situation in Mali." PSC/PR/COMM.(CCCLXXI), April 25, 2013.

France as Penholder Convinces the Council of the Need for MINUSMA

The UN had already deployed in Mali in early 2013 a special political mission called the United Nations Office in Mali (UNOM) in Bamako only and with few resources, based on an authorization by Security Council Resolution 2085 (December 20, 2012) given at the same time AFISMA was authorized.

UNOM was tasked to "assist the Transitional authorities of Mali in developing a roadmap for the electoral process and national dialogue" and supporting them in expeditiously putting in place "a credible framework for negotiations with all parties in the north of Mali who have cut off all ties to terrorist organizations, notably AQIM and associated groups including MUJWA, and who recognize, without conditions, the unity and territorial integrity of the Malian State, and with a view to addressing the long-standing concerns of communities in the north of Mali."[18]

France's strong involvement both militarily (via operation *Serval*) and politically, thanks to its permanent seat on the Council and its role as penholder, contributed to the UN's choice of a peacekeeping operation rather than a UNOM-type "special political mission", or an AMISOM-like African operation that the UN would have provided support to.

In such a context, the option of a political mission was not perceived by the Council as sufficiently robust to restore stability, even if some, notably in the UN Department of Political Affairs (DPA), defended the idea, questioning whether there was really "a peace to keep" in Mali (since there was no peace agreement at the time of the discussion), and the ability of a peacekeeping operation to operate effectively in a terrorist environment alongside an counterterrorist force.

But the United States was not keen on the option of an African operation operating in parallel to a small UN political mission without troops. The US was wary of the UN having to finance such an African operation without the Security Council really having control over it.[19] In Somalia, the Security Council had initially agreed in principle that the UN would provide AMISOM's African contingents with logistical support in exchange for not creating a UN peacekeeping operation. However, this

[18] Security Council, Resolution 2085, S/RES/2085 (2012), December 20, 2012.

[19] Colum Lynch, "Susan Rice floats U.N. peacekeeping force in Mali.", Foreign Policy, January 23, 2013.

support became more and more substantial over the years (rations, fuel, water, housing, infrastructure and maintenance services; while the European Union financed the salaries of African troops[20]), without producing convincing results on the ground.

A UN peacekeeping operation mandated by the Security Council simply offered better guarantees in terms of logistical and financial resources to support military contingents, and better fiduciary control. It also offered a more multi-dimensional presence, including civilian personnel, essential to addressing some of the root causes of the 2012 complex crisis of the country's political, security and socio-economic governance.

France as the penholder on Mali was quick to push for a multidimensional UN peacekeeping mission in Mali, rather than continuing the political mission already in place. A peacekeeping operation with blue helmets seemed a better way of "multilateralizing" its military intervention, and could ultimately provide an exit strategy, even if the blue helmets were to face operational challenges and a complex terrorist threat in Mali. Although an African force would have had a freer hand to serve as an auxiliary to *Serval*, the AFISMA experience had been very mixed (apart from the Chadian contingent, the only one to really fight alongside French soldiers) and France knew it had more influence at the UN than at the AU.

Under-Secretary-General Hervé Ladsous, the Head of the Department of Peacekeeping Operations (DPKO) at the time—who previously chief of staff to French foreign ministers Michèle Alliot-Marie and Alain Juppé—was sometimes accused of "playing into France's hands" at the United Nations, more so than his predecessors, notably for having supported the creation of the two new UN operations in the Central African Republic and Mali both following French military interventions.[21] Hervé Ladsous, at a French Senate hearing in 2016 actually spoke of a specific mode of intervention for France at the United Nations:

> France has a specific mode of intervention at the United Nations. These are often national interventions, but under the cover of a Security Council mandate, coordinated with United Nations activities. It's a formula that has

[20] Walter Lotze and Paul D. Williams, "The Surge to Stabilize: Lessons for the UN from the AU's Experience in Somalia" New York: International Peace Institute, May 2016.

[21] Ibid.

proved to work well, and we'll continue to work in this spirit ... a UN operation is not and is not meant to be counterterrorist tool.[22]

Inspired by the "*clear, hold, build*" (COIN) counter-insurgency strategy developed by the United States in Iraq and Afghanistan, some decision-makers in Paris may have initially thought that Blue Helmets would play a political stabilization role while the French army focused on military tasks. Michael Shurkin, a former analyst at the RAND Corporation, sees in this the legacy of practices and doctrines derived from the French army's colonial and counter-insurgency operations in the days of Generals Gallieni and Lyautey, which concluded to the need for a population-centric focus and a reduced emphasis on military operations:

> Operation *Barkhane* which began in 2014, reflects this new doctrine, which means that the French military focuses on security while waiting for others to do the political work. This is complicated by the fact that the French presence constitutes a political intervention, even as the French try to avoid interfering in politics.[23]

France's permanent representative to the UN at the time of the discussions on the creation of MINUSMA, Gérard Araud, later admitted in his memoirs, not without a certain cynicism, that he had doubts about the relevance of such a UN mission in such a context:

> I had strong doubts about the relevance of sending [to Mali] contingents that I knew would have neither the means nor even the will to fulfill the ambitious mandate I had the Security Council adopt. I actually only overcame the objections from my colleagues [the other ambassadors to the UN] by promising the support of French forces if things went wrong.[24]

Russia which presided over the Security Council the month of the negotiations over the creation of MINUSMA showed willingness to discuss a UN peace operation in Mali but not without expressing concerns:

[22] Senate, "Comptes Rendus de la Commission des Affaires Étrangères, de la Défense et des Forces Armées," March 16, 2016.

[23] Michael Shurkin, "France's War in the Sahel and the Evolution of Counter-Insurgency Doctrine," Texas National Security Review, Vol 4, ISS 1 Winter 2020/2021 | 35–60 https://doi.org/10.26153/tsw/11707.

[24] Gérard Araud, *Passeport Diplomatique*, Paris, Grasset, 2019, p. 81.

> we are disturbed by the growing shift towards the military aspects of United Nations peacekeeping. What was once the exception now threatens to become unacknowledged standard practice, with unpredictable and unclear consequences for the security of United Nations personnel and their international legal status. In that connection, according to the situation clearly delineated in the Secretary-General's report on the situation in Mali (S/2013/189), there should be a clear boundary between peacekeeping and enforcing peace. That is why we believe that MINUSMA's mandate excludes offensive or counter-terrorism operations. Such actions would be counter to the basic principles of peacekeeping.[25]

Pakistan, a major and influential troop contributor and then an elected member of the Security Council along with Russia but also other troop-contributing countries such as Argentina and Guatemala insisted that the basic principles of peacekeeping be reaffirmed in the text of the resolution. They were concerned that MINUSMA would become a counterterrorist mission or another "intervention brigade" like the one approved for the DRC not long before.[26] Pakistan, despite voting in favor of the resolution, refused to send troops to MINUSMA, as did other major troop contributing countries such as India.

Despite the dangerous nature of the operation in Mali, China, usually rather risk-averse, announced in mid-2013 its intention to send a contingent to northern Mali.[27] In so doing, China not only expressed its support to African governments, but also asserted its status as a major power—and the main troop contributing country among the P5—while gradually strengthening and testing its ability to protect its assets and citizens on the African continent, where it has built a strong economic presence.[28]

Resolution 2100 was adopted unanimously by the Security Council on April 15, 2013. It authorizes the deployment of the United Nations Multidimensional Integrated Stabilization Mission in Mali (MINUSMA) from July 1, 2013, with an authorized ceiling of 11,200 military personnel

[25] Security Council Minutes of the 6952nd meeting, S/PV.6952, April 25, 2013.

[26] Security Council Report, "Resolution Establishing a UN Mission in Mali", April 24, 2013.

[27] First elements of 170-strong protection force arrive in Mali in November 2013.

[28] China would later provide financial support to the G5 Sahel joint force in 2017, at the same time as it opened its first logistics base on the continent in Djibouti. See Frans Paul van der Putten, "China's Evolving Role in Peacekeeping and African Security: The Deployment of Chinese Troops for UN Force Protection in Mali," Clingendael Report, September 2015.

and 1440 police officers. Of the two options proposed by the Secretary-General in his report, the Council chose the second, more ambitious option.[29] The Department of Peacekeeping Operations (DPKO) therefore took over from DPA the leadership of the operation in Mali, now known as MINUSMA, which combines the political dimension previously handled by the small United Nations Office in Mali (UNOM) and the military dimension previously handled by AFISMA (managed by the African Union) with the French military operation *Serval* operating in parallel.

A Compromise Between Powers Makes the Mandate Challenging to Implement

Despite the surface consensus when Resolution 2100 was adopted on April 15, 2013 under Chapter VII of the UN Charter, the compromises required for its adoption quickly made its implementation on the ground challenging for MINUSMA. From the very beginning, France the penholder and other member of the Security Council, were hesitant to exert any political pressure on the newly elected Malian government of President Ibrahim Boubacar Keïta (IBK).

The initial mandate includes two main priorities: (1) "Stabilization of key population centres and support for the reestablishment of State authority throughout the country";[30] and (2) "Support for the implementation of the transitional road map, including the national political dialogue and the electoral process." Aware of MINUSMA's capability limits and the dangerous context, the Council also authorized the French troops already on the ground, "within the limits of their capacities and areas of deployment, to use all necessary means, from the commencement of the activities of MINUSMA until the end of MINUSMA's mandate as authorized in this resolution, to intervene in support of elements of MINUSMA

[29] Option 1: Integrated, multidimensional United Nations political presence alongside an African-led military force. Option 2: Integrated, multidimensional United Nations stabilization mission established under Chapter VII of the Charter of the United Nations, a complement to a parallel force. Report of the Secretary-General on the situation in Mali S/2013/189, March 26, 2013.

[30] Including helping to re-establish state administration throughout the country, rebuilding Mali's security sector, agreeing and implementing disarmament, demobilization and reintegration (DDR) programs for ex-combatants and dismantling militias and vigilante groups, etc.

when under imminent and serious threat upon request of the Secretary-General."[31]

The UN was tasked with using its good offices and confidence-building measures to "facilitate progress towards an inclusive national dialogue and reconciliation process" inclusive of armed groups (paragraph 4). As in most contemporary multidimensional peacekeeping operations, these two priorities are complemented by the protection of civilians and UN personnel, the promotion and defense of human rights, support for humanitarian action, action in favor of national and international justice. In addition to these traditional tasks, MINUSMA was the first ever peacekeeping operation to be mandated to also provide "support for cultural preservation"[32] in collaboration with UNESCO, because some of the cultural and historical sites in Mali has been directly targeted by jihadi groups during the occupation of north Mali.[33]

The almost schizophrenic dimension of this initial mandate stems from the fact that some members of the Security Council were more concerned with guaranteeing respect for the sovereignty of the host state (even if that state has "failed"), while others insisted on the need for a politically negotiated solution between the belligerents with a view to establishing a more lasting peace. MINUSMA thus found itself in the difficult position of on the one hand, having to support the Malian state, and on the other hand, having to facilitate a negotiation process between this same state, a party to the conflict, and armed rebel groups.

The political process underpinning MINUSMA was fragile, and for good reasons. There was no political process when the mission was initially conceived in March 2013 in the wake of the French *Serval* military intervention. Just a few days before MINUSMA's official deployment on July 1, in June 2013, Mali's interim government signed a preliminary "agreement for the presidential election and inclusive peace negotiations in Mali"[34] with two Tuareg armed groups—the MNLA and the *Haut Conseil*

[31] Security Council Minutes of the 6952nd meeting, S/PV.6952, April 25, 2013.

[32] Security Council Resolution 2100, S/RES/2100 (2013), April 25, 2013.

[33] See Mathilde Leloup (2019) Heritage Protection as Stabilization, the Emergence of a New 'Mandated Task' for UN Peace Operations, International Peacekeeping, 26:4, 408–430, DOI: https://doi.org/10.1080/13533312.2019.1613894.

[34] Article 24 would allow armed groups that had not participated in the negotiations to join the agreement, including that of the *Mouvement arabe de l'Azawad* (MAA) and the self-defense militia known as the *Coordination des mouvements et forces patriotiques de résistance* (CMFPR).

pour l'Unité de l'Azawad.[35] This preliminary Ouagadougou agreement includes a ceasefire and paved the way for the holding of presidential elections in July and August 2013, bringing the transition period to an end. Article 21 stipulates that, within 60 days of the installation of the new government, the elected Malian government must begin inclusive peace talks with the armed groups that signed or adhered to the agreement, as well as with all the communities of northern Mali.

The same Malian transitional government also formally expressed its consent to the deployment of a UN peacekeeping operation in a letter addressed to the UN Secretary-General and transmitted to the Security Council[36] but expressed a preference for the deployment of UN peacekeepers in northern Mali and a minimal UN presence in the capital Bamako.

However, the strong mandate desired and obtained by the UN operation[37] soon came against the fact that the newly elected Malian President Ibrahim Boubacar Keïta (IBK)—and therefore the host state—was more interested in the UN supporting the redeployment of Malian defense and security forces to northern Mali and the disarmament of armed groups, than in MINUSMA's mandate second priority of supporting peace negotiations with these same armed groups. President IBK and his government launched a number of initiatives, notably the *États généraux de la décentralisation* (decentralization forum) and a *Forum de concertation des communautés du Nord du Mali* (consultation forum with communities of northern Mali) but did not really include armed groups, nor did they abide by the terms of the Ouagadougou agreement signed by their predecessors, notably article 20 which stipulated:

> Following the presidential elections and 60 days after its installation, the new Government, in collaboration with the Dialogue and Reconciliation Commission will begin, with the support of the international community, peace talks with all the communities of the north, the signatories and as well

[35] The HCUA was created in May 2013 from a dissident faction of *Ansar Dine*, the *Mouvement Islamique de l'Azawad*, itself created in January 2013.

[36] Security Council, Letter dated April 16, 2013 from the Secretary-General addressed to the President of the Security Council, S/2013/230, April 16, 2013.

[37] Notably in the Ouagadougou Agreement follow-up mechanisms set up. These are the *Comité de Suivi et d'Evaluation* (CSE) and the Commission Technique Mixte de Sécurité (CTMS), both chaired by MINUSMA, respectively by Special Representative of the Secretary-General and Head of Mission Bert Koenders and by Force Commander General Jean Bosco Kazura.

> as the armed groups that have adhered to the present agreement with a view to achieving the establishment of a comprehensive and definitive peace.[38]

President IBK, who had been officially elected President of Mali with 77.61 percent of the votes casted, did not want to put himself on the same level as so-called rebel armed groups. He would even refuse to meet the Special Representative of the Secretary-General and head of mission Bert Koenders (but did meet with his deputy, Abdoulaye Bathily, a Senegalese professor and politician). He also expressed his dissatisfaction with the presence of white MINUSMA 4 × 4 vehicles in the capital Bamako, which gave the impression of a failed or at least fragile country. He did not appreciate either the fact that MINUSMA set up its headquarters in the most central of Bamako's grand hotels, the *LAICO l'Amitié*, which golf course, tennis courts and large swimming pool were once the favorites of Bamako's elite.

This footprint gave MINUSMA a high profile in the capital, when in fact more than 90 percent of its military personnel were based in northern Mali, in the Gao, Timbuktu and Kidal regions. But the vast majority of mission's civilian employees were indeed based in Bamako, mainly due to the lack of suitable infrastructure in the MINUSMA camps in northern Mali, at least initially. Until the end of 2013, for example, there was only one single-pit latrine in the Kidal camp, and there was no canteen so civilian employees had to bring their food with them.

Mistrust between the Malian government and the armed groups ran deep, and MINUSMA was unable to get the parties to make progress on confidence-building measures such as the release of prisoners. As a result, from October 2013 onwards, the parties stopped meeting in the Ouagadougou Agreement's Monitoring and Evaluation follow-up Committee (CSE), due to a lack of political will on both sides.

This didn't prevent some modest progress, including on October 29, 2013, the government releasing eleven prisoners and lifting arrest warrants against four former parliamentarians who were members of the MNLA and HCUA; and on the other hand, the MNLA vacating the governorate building and the national radio station in Kidal on November 14, 2013, following intense diplomatic efforts by MINUSMA, the African Union, ECOWAS and the European Union. In addition, the MNLA and

[38] Ouagadougou Agreement (between the Republic of Mali and the armed rebel groups MNLA and HCUA), June 18, 2013.

HCUA began to pre-canton some of their fighters at three sites in the Kidal region, with the support of MINUSMA, in preparation for the future Disarmament, Demobilization and Reintegration (DDR) process that should be negotiated as part of a final peace agreement.[39]

Building on the momentum created by a Security Council visit to Mali in early February 2014, during which all Malian players reaffirmed to Council members their willingness to take part in an open dialogue to achieve a comprehensive settlement of the crisis, MINUSMA attempted to revive the stalled political dialogue by organizing a series of informal workshops in February and March 2014 under the auspices of the CSE (and attended by the Minister of Foreign Affairs, the Minister of Defense, the international community and representatives of armed groups and civil society). The main tangible outcome of these workshops was a modest February 18, 2014 Modus Operandi on cantonment. But the overall political process remained stalled.

Instead, the Malian government continued to put pressure on MINUSMA to help redeploy its army and administration to northern Mali (without negotiating with armed groups or consulting communities which often saw the Malian army as predatory rather than bringing security) rather than trying to redeploy (as suggested by the UN) basic services that could have helped improve acceptance of the return of state presence in these regions.

With the lack of progress on the political front, tensions continue to mount, culminating in clashes between the Malian army and armed rebel groups on May 17, 2014, on the occasion of a visit to Kidal by Malian Prime Minister Moussa Mara. Each side blamed the other for triggering the fighting, responsible for the deaths of six civilian civil servants at the Governorate headquarters. The Prime Minister described the events as "a declaration of war [...] which leaves him no choice but to retaliate."[40] Four days later, after sending troops and heavy equipment[41] to the north, Malian forces launched an assault on the town of Kidal. After a few hours of fighting, however, they had to retreat after facing strong resistance from MNLA, HCUA and affiliated armed groups. With 33 soldiers killed, the

[39] Report of the Secretary-General on the situation in Mali S/2014/1, January 2, 2014, paragraphs 5 and 6.

[40] Office of the Prime Minister of Mali, *Communiqué du Gouvernement relatif à la situation à Kidal*, May 17, 2014.

[41] Including two BM-21s or "Stalin's organ", Soviet military trucks with 122 mm multiple rocket launchers.

toll was heavy for Malian troops, who took refuge in MINUSMA camps in Kidal and other towns in northern Mali. These Kidal events marked a decisive turning point in the peace process.[42]

Despite the signing of a ceasefire on May 23[43] committing the signatories to return to the provisions set out in the Ouagadougou preliminary agreement and to immediately resume negotiations, with the support of regional and international partners, the battle of Kidal radically changing the situation—and the balance of power—on the ground. With the Malian administration and army forced out, armed groups now controlled a large part of northern Mali and started to establish a parallel administration including "local security committees." This raised questions within MINUSMA on how to work in these regions with armed groups—that had become "de facto authorities"—without legitimizing them.

At the same time, jihadi groups, some of whom took part in the fighting in Kidal, had begun to reorganize and to target UN peacekeepers. MINUSMA was now the main force operating in northern Mali following the departure of Malian forces[44] and the downsizing and reorganization of French forces with the replacement of Operation *Serval* by Operation *Barkhane* (now a regional military presence extended to four other Sahel countries, Mauritania, Burkina Faso, Niger and Chad) in the summer of 2014.[45]

The Security Council never really clarified which aspects of the mandate—resulting from a compromise between powers—to prioritize or how to implement it on the ground in this new context. Interpretation of the mandate was therefore primarily the responsibility of the Special Representative of the Secretary-General and head of MINUSMA on the ground. The mission often found itself caught between communities who criticized the UN for doing too little, armed groups who suspected the UN mission of being biased toward the government, and a host

[42] Report of the Secretary-General on the situation in Mali S/2014/403, June 9, 2014.

[43] Thanks to the mediation of Mauritanian President and current AU Chairman Mohamed Ould Abdel Aziz and the Special Representative of the Secretary-General.

[44] In practice, the few Malian soldiers who remained in Ménaka and Tessalit after May 2014 remained in their camps under the protection of international forces and raely ventured out.

[45] France 24, "Serval ends in Mali launch of operation "Barkhane" in the Sahel," July 13, 2014, https://www.france24.com/fr/20140713-operation-Serval-mali-france-barkhane-le-drian-defense-militaire-terroriste.

government suspicious of the UN and its representatives, particularly those based in the North and interacting with armed groups leaders on a regular basis.

The host state was particularly concerned that the UN mission should give legitimacy to armed groups. This limited MINUSMA's ability to implement projects (such as the restoration of basic services, water and electricity) in the Kidal region. UN aircraft nonetheless continued to be used for medical evacuations and to transport government officials and other official delegations to and from Kidal.[46]

With MINUSMA struggling to establish itself as an impartial mediator accepted by both the Malian government and the armed groups, Algeria, which had already played a key role in mediating previous Malian crises since the 1990s, once again took the lead in the peace process, albeit initially without a real mandate.

In January 2014, Algiers started "exploratory discussions" with the armed movements of northern Mali in an attempt to bring them together in a coherent platform ahead of negotiations with the government of Mali. These efforts intensified after the reelection of Algeria's President Abdelaziz Bouteflika in April and the clashes in Kidal in May. The efforts only partially succeeded. Instead of bringing all Malian armed groups together in a single coalition, Algiers settled for two separate platforms of armed groups.[47]

At the same time, Algeria's eternal regional rival, Morocco and the King Mohammed VI, received an MNLA delegation in February, led by its Secretary-General Bilal ag Cherif, who was concerned about Algeria's overly prominent role in the upcoming peace negotiations. Algeria was

[46] For example, a government delegation visited Kidal in December 2014 to offer its condolences following the death of the traditional Tuareg leader of the Ifoghas tribe (the Amenokal), and various delegations from armed groups regularly traveled as part of peace negotiations.

[47] On the one hand, the *Coordination des Mouvements de l'Azawad* (CMA) will bring together the MNLA, the HCUA and a branch of the MAA (this alliance already existed since November 2013 and brings together the so-called independence groups that took up arms against the Malian government in 2012). On the other hand, the *Platform* was created on June 14, 2014 to bring together all the other groups closer to the Malian government and with no real military existence on the ground at that stage, at least until the creation of the *Groupe d'autodéfense Touareg Imghad et alliés* - GATIA - in August 2014, which gave the Platform considerable military power. The *Platform* includes the other branch of the MAA, the sedentary and self-defense groups CMFPR and the Coalition du peuple pour l'Azawad (CPA), which had split from the MNLA on March 18 of the same year.

jealous of the fact that its neighbor was hosting the inter-Libyan peace talks under the aegis of the United Nations Mission in Libya (UNMIL) in Skhirat, Morocco.[48] This regional rivalry between Algeria and Morocco continued throughout the peace process, as evidenced by the visits in early 2021 of Malian President N'Daw to Algiers and of his vice-president, Colonel Assimi Goïta, to Rabat, testifying to the balance that Mali's transitional government is trying to strike between these two neighbors.[49] The press even speculated that Algeria could send soldiers from its *Armée nationale populaire* (ANP) under blue helmets to Mali,[50] but this never happened.[51]

In 2014, Algeria's role was legitimized by a joint declaration issued in Algiers by the Malian government, the CMA and the *Platform*, which officially attributed the role of "lead mediator" to the Algerian government, thus marginalizing the previous mediation efforts of ECOWAS and, to a certain extent, MINUSMA. This despite the mandate MINUSMA had from the Security Council and the central role the Ouagadougou Preliminary Agreement had given to the UN. The new process also effectively dropped one of the main stipulations of the Ouagadougou Preliminary Agreement, with the cantonment of armed groups no longer a precondition for dialogue.

But the real novelty of this process compared to past peace processes in Mali was that it involved numerous mediators with diverse interests and with different links to the various parties: the lead mediator, Foreign Minister Ramtane Lamamra of Algeria (who had served as the AU commissioner for peace and security from 2008 to 2013), invited a number of other "co-mediators," including the UN, AU, ECOWAS, EU, Organisation of Islamic Cooperation, Burkina Faso, Chad, Mauritania, Niger, and Nigeria. Apart from those international actors already involved

[48] United Nations, "A new round of inter-Libyan dialogue has begun in Morocco, as the process enters its final phase - press releases." August 27, 2015, https://news.un.org/fr/story/2015/08/317422-un-nouveau-cycle-de-dialogue-inter-libyen-commence-au-maroc-alors-que-le.

[49] Mohammed Jaabouk, "Maliun enjeu géostratégique pour le Maroc et l'Algérie," Yabiladi, March 16, 2021, https://www.yabiladi.com/articles/details/107247/mali-enjeu-geostrategique-pour-maroc.html.

[50] Since the adoption of the new Algerian constitution opens up the possibility for the country to take part in an operation outside its borders for the first time.

[51] Pascal Airault, "What if Algeria were to join the UN military force in northern Mali?..." L'Opinion, April 8, 2021, https://www.lopinion.fr/edition/international/si-l-algerie-integrait-force-militaire-onusienne-nord-mali-241141.

in the implementation of the Ouagadougou Preliminary Agreement, these included neighboring countries with a stake in the outcome of the negotiations, most of which hosted their own Tuareg populations. This "college of mediators" first met in Algiers in July 2014 ahead of the opening of the first phase of the peace negotiations. The delegations of the government of Mali, the CMA, and the Platform were composed of about thirty persons each. The government and CMA brought no more than three women each, and the Platform had an all-male delegation.

After eight months and five phases of negotiations, the Algerian-led mediation team suddenly ended negotiations at the end of February 2015 and invited the parties to initial an agreement it presented as an "acceptable compromise" for all.[52] The lead mediator stressed that negotiations on this document had ended, but that its initialing would pave the way for continued discussions during the implementation phase. Algeria also argued that it was urgent to initial an agreement to counter would-be spoilers in a regional context of terrorism and organized crime. While the parties seemed caught by surprise, the general view among members of the mediation team was that negotiations in Algiers needed to come to an end and that further discussions were not likely to make headway in bridging the positions of the parties.

The government and the *Platform* initialed the document on March 1, 2015, in Algiers as the *Agreement for Peace and Reconciliation in Mali*. But the government delegation produced a twelve-page document listing a number of remaining preoccupations. In particular, the government continued to oppose a single economic zone for northern Mali, the option for regions to join together of their own volition, and any preconditions for the redeployment of the Malian army, including joint patrols and integration of ex-rebels. Foreign Minister Abdoulaye Diop also requested the removal of a reference to "Azawad" in the preamble of the document.

The CMA however refused to initial the document, considering it did not take into account some of its main aspirations, including the recognition of the northern territory it refers to as "*Azawad*" as a geographical and political entity and security arrangements within this territory. It also expressed doubts as to whether the international community could

[52] Arthur Boutellis and Marie-Joëlle Zahar, "Un processus en quête de paix: Les enseignements tirés de l'accord intermalien", International Peace Institute, New York, January 2018.

provide the necessary guarantees for implementation of the agreement and asked for additional time to consult its constituency before initialing it. It however reaffirmed its commitment to respecting the various ceasefire and cessation of hostilities in various communiqués, partly to avoid "targeted sanctions against those who resume hostilities and violate the ceasefire," a threat already brandished by certain members of the Security Council.[53]

In his address to the Security Council on April 9, 2015, Malian Foreign Minister Abdoulaye Diop stated that his government's position was that "the negotiations are over", that progress had to be made in implementing the agreement, and called on the "international community" to "isolate [...] and impose sanctions" against those "who will not sign the agreement."[54]

Even though MINUSMA lost its lead on political negotiations to Algeria, the UN mission has managed to maintain its leadership position on security issues, and for good reasons, as it has almost 12,000 blue helmets on the ground in Mali. Each phase of negotiations was accompanied on the ground by clashes and heightened tensions between Malian forces, the *Platform* and the CMA, as armed groups fragmented, and alliances recomposed according to the balance of power in anticipation of a possible agreement. The challenge for MINUSMA was to maintain the perception of the impartiality of its peacekeepers, while acting and working closely with other members of the international mediation team who themselves were not necessarily impartial or perceived as such. Neighboring states were notably concerned about the possible domestic repercussions of autonomy or even federalism in northern Mali.

It was only under strong international pressure, particularly from France and the United States—who had joined the Algiers process as observers toward the end of the process—that the CMA finally agreed to initial the agreement in Algiers on May 14, 2015, on condition that direct discussions on its demands took place with the Malian government before the final signing ceremony. After the *Platform* took military control of Ménaka on April 27, the CMA also made its adherence to the peace

[53] Statement by the President of the Security Council S/PRST/2015/5, February 6, 2015.

[54] Mali's Minister of Foreign Affairs Abdoulaye Diop, Press conference after the Security Council meeting in 2015, June 23, 2015, https://www.un.int/mali/mali/déclarations-de-la-délégation-du-mali-devant-le-conseil-de-sécurité-en-2015.

agreement conditional on a withdrawal of *Platform* forces occupying Ménaka.[55] This is also why the CMA refused to join the Malian government and the *Platform* for the signing ceremony organized in Bamako on May 15, 2015.

The UN Secretary-General's statement delivered at the ceremony by the Under-Secretary-General for Peacekeeping Operations, Hervé Ladsous, recommended that the agreement should remain open for signature by the remaining parties, and that dialogue continue. It also advised against using the signing ceremony as a pretext for relaunching military operations against non-signatory groups. This message from the UN was poorly received by Mali's President IBK, who in turn accused MINUSMA of partiality, thus highlighting major differences with the host government regarding the future of the peace process after the signing of the agreement.[56]

On June 5, 2015, following further consultations on implementation modalities, (Fig. 2.1) the CMA finally announced that it would sign the peace agreement, following a deal between the CMA and the Government regarding the withdrawal of *Platform* forces occupying Ménaka, and the promise of direct discussions with the Malian government. On June 19, peacekeepers temporarily took over security of Ménaka and its population in return for a halt to CMA attacks in northern Mali. On the same day, the Malian government lifted the existing arrest warrants against 15 prominent CMA leaders. Finally, on June 20, Sidi Brahim Ould Sidatt signed the "Agreement for peace and reconciliation in Mali resulting from the Algiers process" in Bamako on behalf of the CMA, in the presence of over 150 CMA representatives from Kidal and the sub-region.[57]

[55] The Platform forces that invested Ménaka were commanded by Yoro Ould Daha, who had been commissioner of the MUJAO's Islamic police in Gao during the Islamists' occupation of northern Mali by the Islamists in 2012. See Radio France Internationale (RFI), "Mali: Qui tient la localité de Ménaka?", May 17, 2015, www.rfi.fr/afrique/20150517-mali-tient-localite-menaka-yoro-azawad-casques-bleus-groupes-armes-cma.

[56] International Crisis Group, "MaliLa paix à marche forcée?", May 22, 2015.

[57] Agreement for Peace and Reconciliation in Mali signed in Bamako on June 20, 2015. Full French text available at http://peacemaker.un.org/node/2681.

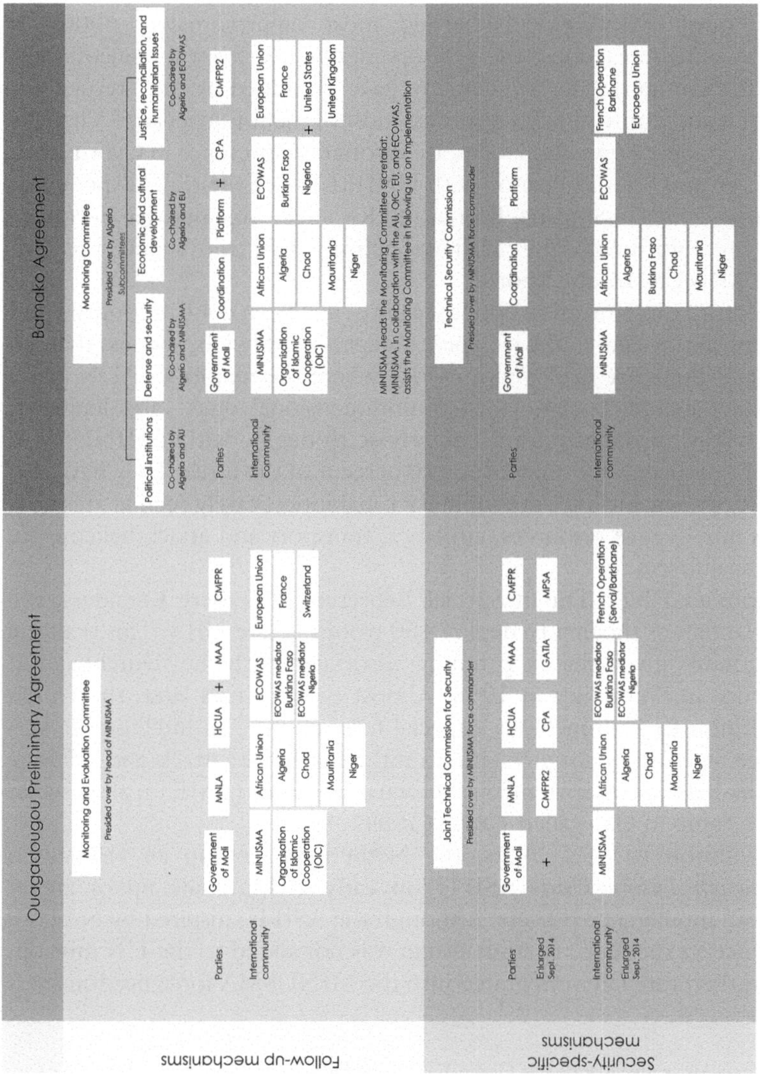

Fig. 2.1 Implementation and follow-up mechanisms to the different peace agreements

The So-Called Return of Western Troop Contributing Countries

Even though MINUSMA inherited most (approximately 6000) of AFISMA's African troops, the UN operation was, from its inception in 2013, the laboratory for what many have described as the "return" of Western troop-contributing countries, most of them European.[58] Virtually absent from peace operations (despite an ongoing presence in Cyprus with UNFICYP and South Lebanon with UNIFIL) since their UN experiences in the mid-1990s in Somalia, Bosnia and Rwanda, some of these European contributors got involved in Mali and actively tried to influence the conduct of MINUSMA operations.

In recent years European countries have only been contributing around 8 percent of the total number of peacekeepers worldwide, and less than 40 percent of them are deployed in Africa where the majority of UN peace operation—including the largest multidimensional ones—are deployed. Nevertheless, most European states chose to deploy within MINUSMA, representing over 1000 soldiers, or 10 percent of the force. They have also provided significant high-end military capabilities, rarely seen at the UN, such as intelligence resources, airplanes, transport and attack helicopters, and special forces.

Encouraged by a Dutch Special Representative, Bert Koenders, the Netherlands was the first to deploy 450 troops in late 2013. This is all the more notable given the deep trauma associated with the 'Dutchbat III' experience in Srebrenica in 1995. Almost twenty years later, the Dutch contribution to Mali includes a special forces unit,[59] an intelligence, surveillance and reconnaissance (ISR) unit, and *Apache* attack and *Chinook* transport helicopters with medical evacuation teams (AMET), all based at a Dutch camp in Gao called *Camp Castor*.

In addition to these units, the Netherlands set up an *All-Sources Information Fusion Unit* (ASIFU) in early 2014, made up of eighty European intelligence officers. Although the ASIFU, inspired by NATO's experience s experience in Afghanistan was not suited to the UN mission, and was therefore later integrated into the MINUSMA force headquarters

[58] Joachim Alexander Koops and Giulia Tercovich, *European approaches to United Nations peacekeeping: towards a stronger re-engagement?* (London, 2018).

[59] including some 20 Danes added after the appointment of Danish general Michael Lollesgaard to command the MINUSMA force in 2015.

(as part of the "U2" military intelligence), it did contribute significantly to advancing the taboo debate on intelligence at the UN.[60] The last Dutch unit (an LRRP long-range reconnaissance patrol) left Mali at the end of 2019, but the Netherlands continued contributing a C-130 military transport aircraft to the mission in the form of rotations with other European countries.[61]

Sweden also deployed an ISR unit to Mali in 2014 in Timbuktu, comprising around 220 personnel, making it the only European troop contributing country to have sent personnel to MINUSMA beyond Gao (where the largest French military base was located and therefore providing some security cover for the UN) and Bamako. This deployment, initially scheduled to last eighteen months, ultimately lasted five years. After leaving Timbuktu, Sweden joined the other Europeans in Gao with a contingent of 200 peacekeepers, which was due to remain until 2024 (but the Swedish Parliament decided to recall the contingent a year earlier than planned as relations with the host country were straining).[62]

In 2016, Germany took over from the Dutch in Gao, initially deploying 600 soldiers, including *Tiger* attack helicopters and NH90 transport helicopters as well as an ISR unit; it hosted two NH90 helicopters and a Belgian ISR section for a short time in 2018 (while Belgian General Jean-Paul Deconinck commanded the MINUSMA force), an Irish special forces section, and a Lithuanian protection unit in 2017. In May 2022, Germany surprised many experts by announcing that it was not only extending its participation in MINUSMA, but also reinforcing it, increasing its troop strength from 1100 to 1400 soldiers.[63] This despite France withdrawing from Mali and MINUSMA lacking combat helicopters, and after twelve

[60] See Alexandra Novosseloff and Olga Abilova. "Demystifying Intelligence in UN Peace Operations: Toward an Organizational Doctrine." International Peace Institute, May 9, 2020, https://www.ipinst.org/2016/07/demystifying-intelligence-in-un-peace-ops and Sarah-Myriam Martin-Brûlé, "Finding the UN Way on Peacekeeping-Intelligence." International Peace Institute, April 11, 2020. https://www.ipinst.org/2020/04/finding-the-un-way-on-peackeeping-intelligence.

[61] Arthur Boutellis and John Karlsrud, "Plug and Play: Multinational Rotation Contributions for UN Peacekeeping Operations." NUPI/IPI, 2017.

[62] AFP, "Mali: Sweden quitter la Minusma un an tôt que prévu", March 4, 2022.

[63] RFI, "Mali Germany veut prolonger et renforcer ses effectifs à la Minusma", May 11, 2022.

German peacekeepers had been injured in a suicide vehicle attack at the end of June 2021.[64]

In the summer of 2018, Canada replaced the German helicopter component for just over a year (with three *Chinooks*, five *Griffins* and 250 personnel, including three AMETs), followed by Romania with four transport Puma helicopters (including two AMETs) in October 2019. The Romanian helicopters, whose preparation was made possible by US support and the deployment facilitated by Canada, also benefited from the infrastructure and support of the German camp. When the latter left Mali at the end of 2020, the UN had not found another contributing country to replace the helicopter unit.

The last Western country to join MINUSMA was the United Kingdom which deployed a 250-strong Long Range Reconnaissance Patrol (LRRP) in 2021 to replace the Dutch unit. This deployment came against a backdrop of *Brexit* and the need for Britain to reaffirm its international legitimacy linked to its status as a permanent member of the Security Council. The UK had already operated in Mali as part of the French *Barkhane* operation with *Chinook* helicopters since 2018. The UK had also previously contributed to other UN operations in Africa, sending 400 troops from an engineering unit and a level II military hospital to the mission in South Sudan (UNMISS) from 2017 to 2020.[65]

The British troops quickly distinguished themselves by their ability to operate for long periods away from their base (concluding a 28-day patrol[66]) and their proactive attitude (seizing weapons from armed groups affiliated to ISGS[67]), and by their ability to communicate about their actions. In October 2021, British peacekeepers killed two armed men who had opened fire on their vehicles after being chased. Lieutenant-Colonel Will Meddings, commander of the LRRP, told the press:

> Today's action demonstrates exactly what the UK is bringing to the UN's most dangerous peacekeeping mission, a long range force that doesn't just

[64] France 24, "Mali several soldiers killed and peacekeepers wounded in two separate attacks," June 25, 2021.

[65] The only other major European contribution to a peace operation in Africa during the same period was Portugal's contribution to MINUSCA in the Central African Republic, in the form of a rapid reaction force of 160 paratroopers and commandos, which has demonstrated its effectiveness time and again since 2017, in terms of intelligence gathering and not hesitating to use force if necessary, unlike many other UN contingents.

[66] Janes, "British Peacekeepers Complete First Long-Range Patrol in Mali." Janes.com. April 15, 2021.

[67] Hazel Shearing, UK troops seize IS weapons in Mali operation, May 2021, BBC News.

find those who would harm civilians but acts as well. Results like this come from patrolling huge distances, day and night, in places where Islamic State in the Greater Sahara (ISGS) feel they have the freedom to extort and murder. They cannot act with impunity.[68]

The motivations behind Western countries' commitment to peacekeeping operations are multiple and complex. First and foremost, it is a political decision, based on national interests and considerations of status and values, and a concrete way to demonstrate commitment to multilateralism in general and peacekeeping in particular.[69] As far as Mali is concerned, the Sahel is a strategic priority for the European Union, not only because of the terrorist threat, but also because of illegal trafficking and migrations.

France encouraged other Europeans to become more involved in this region, where it has historical and strategic interests, especially after the terrorist attacks in Paris in 2015. President Macron triggered Article 42.7 of the Lisbon Treaty—a never before used mutual-defense clause. This led Germany to reassess its foreign and defense policy and it influenced its decision to contribute troops to MINUSMA, conceived in part as an effort to support French troops in the Sahel.[70] While some Europeans chose to contribute directly to the French counterterrorist operation *Barkhane* (United Kingdom, Denmark, Estonia, Spain, and Belgium)[71] or to the European *Takuba* Task Force[72] (France, Czech Republic, Estonia and Sweden), most European politicians remained more comfortable in principle with contributing to a UN peacekeeping force, which is less offensive and less risky than an counterterrorist operation.

The former Dutch Defense Minister was also instrumental in lobbying his colleagues to broaden Europe's commitment to MINUSMA. Pressure

[68] Jerome Starkey. "British Troops Kill Two 'Terrorists' in Mali in First Gun Battle in Seven Year." The Sun. October 20, 2021.

[69] See Alex J. Bellamy and Paul D. Williams eds, Providing Peacekeepers: The Politics, Challenges, and Future of United Nations Peacekeeping Contributions (Oxford: Oxford University Press, 2013) and www.providingforpeacekeeping.org.

[70] Reuters, "EU unanimously approves France's request for aid", November 17, 2015.

[71] Ministry of Defence, "BARKHANE: An international training force". July 2, 2020.

[72] The European Takuba Taskforce, composed mainly of special forces, was announced at the summit held in Pau, France on January 13, 2020, and was launched on the ground in Ménaka, Mali in July 2020, with the aim of "complementing" the actions of the French Force Barkhane in support of the Malian Armed Forces (FAMa), in the fight against terrorism in the Sahel states. See France 24, "Swedish Soldiers Join European Takuba Force In Mali." February 5, 2021.

from the United States Obama administration in the context of the September 2015 *Leaders Summit* to make contribution to UN peacekeeping also played an important part.[73] Many countries also contributed to MINUSMA as part of their campaign for a non-permanent seat on the UN Security Council (Article 23 of the Charter). These include the Netherlands, Sweden, Germany, Belgium, Norway, Ireland and Canada, the latter being the only among those that failed to get elected. Once a country has been elected, having peacekeepers on the ground is also a means of more meaningfully contributions and possibly influencing discussions in the Council. For instance, Pakistan, who was an elected member on the Council in 2013, was very active during debates on the creation of MINUSMA.

NATO's withdrawal from Afghanistan has of course been a major factor in European countries' renewed interest in UN operations, in which their military can maintain a degree of (inter)operability. Contributing to peacekeeping can also help national recruitment campaigns and justify the maintenance of budgetary allocations for their armies. Finally, it's an opportunity for these countries to train their troops in difficult theaters, to test new equipment, and sometimes to observe and accompany allied countries (particularly African ones) to which they also offer equipment and training (in Mali, this was the case for Germany, Belgium, and Sweden), as well as to observe other troops from China (based in Gao as part of the peacekeeping force) as well as Russian *Wagner* mercenaries from 2021, who naturally observe them in return.

Unlike other traditional troop contributing countries, Western contributions are not limited to the military capabilities these countries offer to the peace operation, as they are generally accompanied by financial support (notably to the MINUSMA *Trust Fund*) for the implementation of the peace operation's mandate. Germany, for example, financed a $14 million project to rehabilitate the runway at Gao airport and later the airfield in Kidal as well, something MINUSMA had been unable to obtain from the Fifth Committee of the United Nations which votes on mission budgets. Similarly, the Netherlands supported a $4 million project to help integrate the ASIFU into the MINUSMA force headquarters (U2).

[73] The Trump administration (2017–2020) has not pursued this line, asking its allies for better sharing of the "burden" of NATO spending rather than more contributions to UN operations. See Jean-Pierre Stroobants. "At NATO Donald Trump is putting Europeans under pressure and not reassuring them." Le Monde, May 26, 2017.

ASIFU is a good example of the limits of the influence of European countries through this type of "advanced" contribution. In fact, the intelligence fusion model imported from NATO met resistance from the UN, as it neither adapted to the environment (offering strategic intelligence that did not really benefit the contingents on the ground that required more tactical intelligence to improve their security) nor to the UN institution, which has difficulty dealing with secrecy (classification) and suspected ASIFU of producing intelligence "target packs", including for the non-UN French counterterrorist force.[74]

On the political side, however, the participation of Western countries in MINUSMA helped to keep the attention of Brussels and New York on what was happening in Mali and gave the UN operation added political weight and credibility vis-à-vis the host government and armed groups. Although MINUSMA was besieged by visitors, its leadership generally welcomed visits from Western parliamentarians and ministers of defense and foreign affairs, who could then relay messages, particularly when the country was at the same time a permanent member (United Kingdom) or non-permanent member (Germany, Norway, the Netherlands, Sweden, and Belgium) on the Security Council.

A More Coherent Mandate but the Difficult Implementation of an Imposed Peace

Following the signing of the 2015 peace agreement, the Security Council was able to articulate a new, clearer and more coherent mandate for MINUSMA in Resolution 2295 of June 2016:

> Affirms that the gradual restoration and extension of State authority across the territory of Mali, in particular the reformed and reconstituted Malian Defence and Security Forces (MDSF), consistent with the provisions of the Agreement, would contribute significantly to the stability of Mali and to deter the terrorist threat, and, in this regard, encourages bilateral and multilateral partners to increase their support to accelerate the redeployment of the MDSF, once reformed and reconstituted, to the north of Mali, particularly by providing relevant equipment and training, in coordination with the

[74] John Karlsrud and Alexandra Novosseloff, "Doing Less With More? The Difficult 'Return' of Western Troop Contributing Countries to United Nations Peacekeeping", GGI Analysis, Brussels: Global Governance Institute, 2020.

Government of Mali and MINUSMA and within the Framework of the Agreement.[75]

The June 2015 inter-Malian peace agreement admittedly provided political and security solutions to some of the root causes of the conflict, but was not going to solve all the problems, particularly since it signed under international pressure from Algeria, but also France and the United States, which joined the mediation process toward the end as observers.

Because negotiations were cut short by Algeria, the agreement offers a useful framework, but one that remains rather vague. There are, for example, no quotas or numbers to guide the integration of ex-rebels into the reformed and reconstituted Malian Defense and Security Forces (these quotas were only agreed to in the summer of 2022![76]). Mali's history of unimplemented agreements fueled a deep-seated sense of mutual distrust that persisted between the populations of the North and South. The context was marked by an absence of political will on the part of the Malian parties, and while the "international community" has led them to sign peace, it would take time for the Malians themselves to make peace.[77]

Almost a year after the initialing of the Agreement, implementation had not progressed. The signatory parties agreed on a new timetable for the implementation of the main provisions of the Agreement relating to security institutions but the UN Secretary-General expressed concerns in his May 2016 report:

> The obstacles to implementing the Agreement are formidable, and progress to date has been uneven. Political and institutional reforms, particularly those relating to decentralization, the security sector and the rule of law, offer a historic opportunity for a lasting solution to the governance and security issues that have long been at the heart of the Malian crisis. I therefore remain concerned by the slow pace of implementation of the Agreement, and share the sense of urgency expressed by Malian and international stakeholders.[78]

[75] Security Council Resolution 2295, S/RES/2295 (2016), June 29, 2016.

[76] TV5Monde, "Mali: accord pour une intégration de 26.000 ex-rebelles dans l'armée", August 6, 2022.

[77] Arthur Boutellis and Marie-Joëlle Zahar, "Un processus en quête de paix: Les enseignements tirés de l'accord intermalien", International Peace Institute, New York, January 2018.

[78] Report of the Secretary-General on the situation in MaliS/2016/498, May 31, 2016, para. 78.

The international mediation team organized high-level ministerial meetings of the Comité de suivi de l'accord (CSA), in Algiers on January 18, 2016, and again on February 10, 2017, in an attempt to relaunch the implementation process. The adoption of a new roadmap on June 12, 2017 for the implementation of the agreement—beyond the end of the two-year interim period after signature, the appointment of an Independent Observer to objectively assess the state of implementation of the agreement (role taken on by the *Carter* Center),[79] and the ministerial reshuffle of April 11, 2017, accompanied by the activation of a consultation framework between the Government and armed groups, helped to keep the peace agreement alive.

The parties seemed more interested in the personal benefits they derive from the process, such as the monthly allowances for armed groups' representatives on the CSA.[80] But no one wanted to be perceived as obstructing the implementation of the agreement, so as not to expose oneself to possible sanctions[81] and the international community kept pushing for even symbolic advances.

And the implementation of the agreement did progress in 2017. Interim authorities were finally installed in Kidal, Gao and Timbuktu, and new regions created in Ménaka and Taoudéni. On the security front, the Operational Coordination Mechanism (MOC) was established in Gao, and preparations for the disarmament, demobilization and reintegration (DDR) process and mixed patrols were made. The first such mixed patrol took place in Gao on February 23, 2017, despite a particularly deadly suicide attack on the MOC on January 18.[82]

[79] On September 11, 2017, the Comité de Suivi de l'Accord (CSA) finally appointed the Carter Center as an Independent Observer "tasked with objectively assessing the state of implementation" of the agreement, as provided for in Article 63 of the agreement. This independent observer, with a small team on the ground in Mali will publish a "comprehensive report every four months, taking stock of the implementation of commitments made under the terms of the agreement, identifying any bottlenecks, pinpointing responsibilities and making recommendations on measures to be taken".

[80] L'Indicateur du Renouveau, "Pour des per diem non payés: les groupes armés boycottent les travaux de sous-comités du CSA." maliweb.netSeptember 11, 2017. https://www.maliweb.net/la-situation-politique-et-securitaire-au-nord/perdiem-non-payes-groupes-armes-boycottent-travaux-de-comites-csa-2525312.html.

[81] See next section on sanctions.

[82] Célian Macé, "A Gao, le sanglant message d'Al-Qaeda à l'ONU et à la France.", Libération, January 18, 2017.

Ansar Eddine leader Iyad Ag Ghali, one of the main players in the Tuareg rebellion of the 1990s, who has pledged allegiance to al-Qaeda,[83] denounced the peace process as early as October 2015, a month before claiming responsibility for a major terrorist attack on Bamako's *Radisson Blu* hotel.[84] Sometime later, the attack on the Gao MOC killed 64 dead and wounded another 115 wounded among the members of the first mixed patrol (Malian army, CMA and *Platform*). Shortly afterwards, in March 2017, the main Malian jihadist groups[85] united under the banner of *Jamaat Nosrat al-Islam wal-Mouslimin* (JNIM), a coalition led by Iyad Ag Ghali. In the year following the signing of the peace agreement only, Malian defense and security forces suffered 84 attacks, peacekeepers 66 and French forces 15; in the following year, these attacks increased, with 105 attacks against the Malian army, 76 against MINUSMA and 20 against *Barkhane* respectively.[86]

The long-awaited *Conférence d'entente nationale* (CEN), held from March 27 to April 2, 2017, produced a number of recommendations designed to inform the *Charter for Peace, Unity and National Reconciliation* envisaged in the peace agreement. The final report also called for discussions with Malian jihadists, radical preacher Hamadoun Kouffa and Iyad Ag Ghali. The charter presented to President Keïta on June 20 was inspired by the conference's conclusions, but was immediately rejected by the CMA due to the government's decision not to mention *Azawad as a* political and institutional entity. In addition, armed groups accused the Government of acting unilaterally and trying to impose faits accomplis, whether in the appointments of interim authorities, the designation of a new Governor for the Kidal region—a figure reportedly close to the GATIA,[87] or the adoption of the *Code des collectivités territoriales* in the Fall of 2017.

[83] Jules Crétois, "Fusion de groupes jihadistes au Sahel, sous la bannière d'al-Qaïda," Jeune Afrique, March 2, 2017, www.jeuneafrique.com/408773/politique/fusion-de-groupes-jihadistes-sahel-banniere-dal-qaida/.

[84] TV5Monde, "Mali Attaque Armée à L'hôtel Radisson De Bamako." December 24, 2021. https://information.tv5monde.com/info/mali-fusillade-l-hotel-radisson-bamako-68692.

[85] AQIM, *Ansar Eddine*, *Al-Mourabitoune* and the Macina Liberation Front.

[86] Report of the Secretary-General on the situation in MaliS/2017/478, June 6, 2017, para 17.

[87] Report of the Secretary-General on the situation in MaliS/2017/271, March 30, 2017, para. 8.

Military clashes between signatory armed groups continued sporadically, with violent fighting between signatory armed groups in the summer of 2016 and again in the summer of 2017, in violation of the peace agreement. The truce that followed was this time accompanied by local talks between armed groups and the Arab and Touareg communities with which they are affiliated.

Since December 2015, MINUSMA had a new head of mission, Chadian Mahamat Saleh Annadif. His appointed cemented Chad's commitment and growing regional influence from the Lake Chad Basin—where Chadian troops had fought Boko Haram—to the Sahel—where Chadian troops had fought alongside French forces in northern Mali in early 2013 and remained thereafter as part of MINUSMA. Chad had also just served as a non-permanent member of the Security Council in 2014–2015 and had already secured the position of Deputy Force Commander of MINUSMA (General Oumar Bikimo in 2014–2015).

Mahamat Saleh Annadif, a diplomat who had been involved in the rebellion in his own country in the 1980s, immediately imposed a different style. He played an important role in the so-called *Anéfis 2* talks that put an end to the fighting between armed groups in the summer of 2017. The "international community" had initially been reluctant to get involved in the *Anéfis 1* process in October 2015, perceived as a "commercial deal" or "sharing of the pie" between traffickers.[88] From September to November 2017, the new head of MINUSMA launched discreet discussions between armed groups and community leaders that resulted in "commitments" signed by the rebel groups but also by the government and the "international community", including MINUSMA and Algeria. The agreement intended to help resolve underlying tensions between communities and armed groups which had not been addressed in the Algiers talks and to help reduce tensions between traffickers.[89]

While Algeria formally retained the role of leader of the international mediation in the implementation phase of the agreement—and despite sometimes diverging interests within the international mediation team—the new MINUSMA leadership managed to get more involved on the

[88] Crisis Group, "Mali: La Paix Venue D'en Bas?", August 13, 2016. https://www.crisisgroup.org/fr/africa/west-africa/mali/mali-peace-below.

[89] Arthur Boutellis, Delphine Mechoulan, and Marie-Joëlle Zahar, "Parallel Tracks or Connected Pieces? UN Peace Operations, Local Mediation, and Peace Processes," International Peace Institute, December 2020.

ground in Mali in pushing the parties to implement the agreement. Algeria, to the surprise of many within the international mediation, had decided to remain in the lead and notably to chair the *Comité de Suivi de l'Accord* (CSA) during the implementation phase even though MINUSMA had a full-fledged operation in Mali, present from Bamako to Kidal. Algeria however proved reluctant to use pressure and incentives on the parties to achieve progress in implementing the agreement between the monthly meetings of the CSA. In this context, MINUSMA has been able to progressively assume a greater role by leveraging its field presence as well as its role as CSA Secretariat and the SRSG Mahamat Saleh Annadif's active good offices.

Assessing the First Phase: Unanimous Council but Frictions with the Region and Host State

The first phase in the life of MINUSMA took place under the aegis of a Security Council that was unanimous, at least in appearance, on the direction the UN peace operation in Mali should take. The Council followed France's lead, which pushed for the operation in the wake of French military Operation *Serval*, thereby "multilateralizing" its involvement in the country and legitimizing its military intervention. The participation of Chinese and Western contingents, and the support given to MINUSMA by President Obama's US administration, reinforced the Council's interest in the mission, despite the reluctance of some traditional troop-contributing countries to deploy in a context affected by terrorism.

The main frictions were with the African Union (AU) Peace and Security Council, which would have liked to see an African counterterrorism operation continue but financed by UN assessed contributions. The AU would also have liked to see an African appointed to head MINUSMA, rather than the Dutchman Bert Koenders. With the arrival of a new Malian government following the summer 2013 elections but also because of initial mistakes the UN may have made (in terms of footprint and communication), relations with the host government also started to gradually deteriorate. Although Mali state had formally given its consent at the start of the mission, the newly elected government subsequently did not hesitate to marginalize MINUSMA and its leader and reinterpret the mission's mandate to its own preferences.

The visit of Security Council to Mali in early February 2014 was largely complacent toward President IBK and did not help MINUSMA break the deadlock in the political process, nor did it prevent escalation between the belligerents. Tensions rose to the point of direct confrontation between the Malian army and armed rebel groups in May 2014, which reshuffled the political and security cards. Regional players took center stage when a new ceasefire was signed on May 23, 2014. In addition to the presence of the Mauritanian President and the then AU Chairman, Algeria quickly established itself as the "leader of the international mediation", at the expense of ECOWAS and MINUSMA. The AU, whose Peace and Security Commissioner at the time was Algerian Smaïl Chergui, supported Algeria's approach, despite the mandate and prerogatives conferred to MINUSMA by the Security Council since the June 2013 Ouagadougou preliminary agreement.

MINUSMA therefore had to contend with regional powers, but with 13,000 blue helmets on the ground nonetheless retained some leadership on security issues, operating in parallel with the French counterterrorist operation (which became the regional French counterterrorist force *Barkhane* in the summer of 2014). MINUSMA had however become the main armed force operating in northern Mali—after the Malian army was defeated in Kidal—and faced an increasing number of deadly terrorist attacks. The lack of clarity and prioritization in the initial paradoxical mandate to support the re-establishment of state authority and the peace process at the same time, did not make the mission's task easy.

In practice, MINUSMA's leadership retained considerable leeway vis-à-vis the Security Council in interpreting and implementing the mandate. Council member states were not very intrusive, but the Security Council and France—as the penholder with a parallel military force on the ground—did not necessarily help the mission either, refusing in particular to put pressure on the host government to advance the political process that was central to MINUSMA's mandate and critical to the success of the peacekeeping operation (Fig. 2.2).

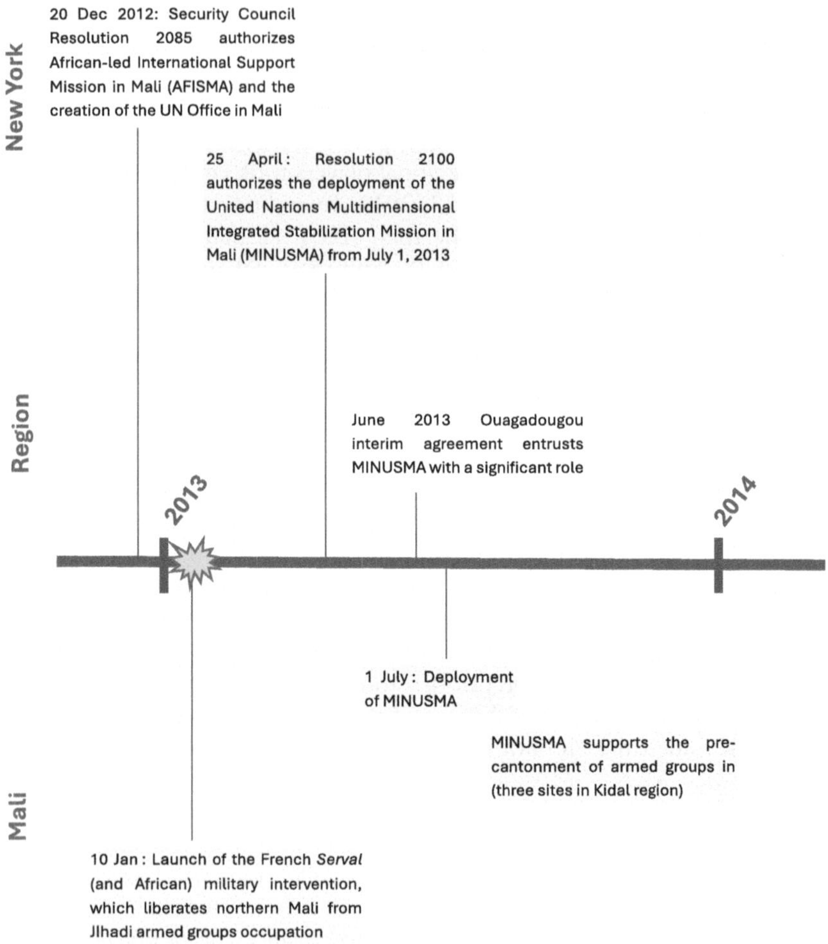

Fig. 2.2 MINUSMA timeline (2013–2015)

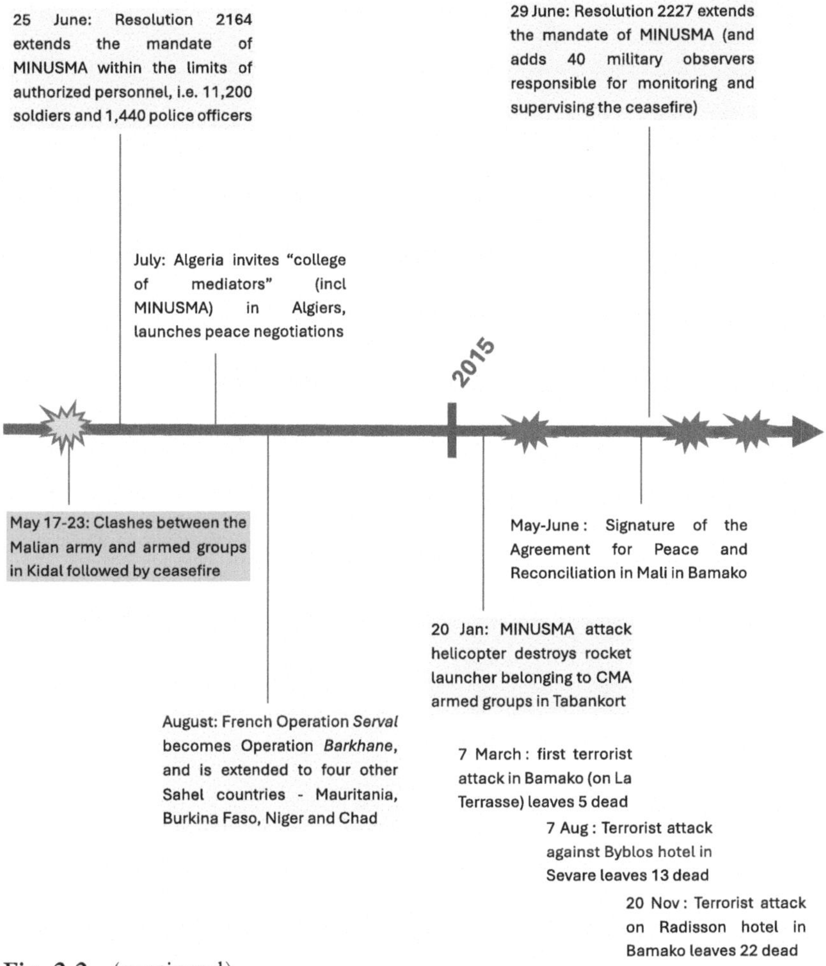

Fig. 2.2 (continued)

CHAPTER 3

MINUSMA Act Two (2016–2019): The Relevance of Peacekeeping in a Terrorist Environment in Question

"In Mali our peacekeeping mission has a mandate adapted to a given political situation that has evolved. It is deployed in an environment polluted by terrorists, who are imposing an asymmetrical war on it. But the fight against terrorism is not part of MINUSMA's mandate whereas we are suffering the consequences. We are not a peace-enforcement mission. But would that really be the solution?"[1]

Mahamat Saleh Annadif, MINUSMA head of mission, 2017

The Adoption of "A More Proactive and Robust Approach" Satisfies the Host State but the Possible Use of Force in a Terrorist Environment Divides the Security Council

The UN mission, with over 10,000 troops and a civilian presence in Bamako, Kidal, Gao and Timbuktu, has generated high expectations and, rightly or wrongly, Malian parties have counted on it to restore security and often criticized it for failing to do so. Indeed, several senior Malian officials suggested already at the time that the mission should make its resources available to the Malian state. One of them even suggested that

[1] *Le Monde*, Mali: "La lutte antiterroriste n'est pas dans le mandat de la Minusma," November 17, 2017.

A. Boutellis, *The UN Stabilization Mission in Mali (MINUSMA)*, Twenty-first Century Perspectives on War, Peace, and Human Conflict, https://doi.org/10.1007/978-3-031-66466-3_3

the UN should authorize the Malian armed forces to call on MINUSMA *Apache* attack helicopters to respond to security issues in the north of the country.

This growing pressure from the host state, but also from certain countries in the sub-region, against a backdrop of increasing terrorist attacks against Malian forces (particularly in the center of the country) and against UN peacekeepers, has led to a questioning of the validity of MINUSMA's "peacekeeping" mandate only Two years after it first deployed to Mali. All of these actors were calling for a more explicit role for the UN mission in the fight against terrorism.

This eventually led to a change in the MINUSMA mandate in June 2016. The renewal of the mandate was accompanied by an increase in the troop ceiling by an additional 2500 peacekeepers. While "the mission's strategic priority is to support the implementation of the Agreement for Peace and Reconciliation in Mali" and MINUSMA is already authorized to "use all necessary means to fulfil its mandate, within the limits of its capabilities and in its areas of deployment" under Chapter VII of the United Nations Charter, the new Resolution 2295 (2016) adopted unanimously by the Security Council now included the following language:

> Requests MINUSMA to move to a more proactive and robust posture to carry out its mandate …
>
> In support of the Malian authorities, to stabilize the key population centres and other areas where civilians are at risk, notably in the North and Centre of Mali, and, in this regard, to enhance early warning, to anticipate, deter and counter threats, including asymmetric threats, and to take robust and active steps to protect civilians, including through active and effective patrolling in areas where civilians are at risk, and to prevent the return of armed elements to those areas, engaging in direct operations pursuant only to serious and credible threats.[2]

This new language was welcomed by Mali's Minister of Foreign Affairs Abdoulaye Diop, who spoke by videoconference from Bamako:

> I thank all Council members for the consensus reached regarding my country. The new MINUSMA mandate incorporates the expectations of the Government of Mali, providing innovations that will enable the Mission to fully accomplish its mandate and assume its role in supporting the imple-

[2] Security Council Resolution 2295, S/RES/2295 (2016), June 29, 2016, para 18 and 19.

mentation of the Agreement on Peace and Reconciliation in Mali and all stabilization efforts within Mali, to which Council members have contributed significantly. We all know that this new strengthened and robust mandate provides the Mission with the means to fulfill its mandate in an environment of exceptional security challenges, characterized by asymmetric attacks.

Mali welcomes the adoption of the resolution, whose strategic objective remains the restoration of State authority throughout our national territory, particularly in the northern regions affected by the crisis. I should like, on behalf of the people and Government of Mali, to extend to all States members of the Security Council our heartfelt gratitude not only for this historic act, but also for the important progress which its adoption represents and above all for sending a strong signal to all the enemies of peace that the United Nations, Mali and the forces of peace will not roll over and will respond in kind in order to continue along the path for peace.[3]

France, which supported Mali's request for a more robust mandate, stressed through the voice of its ambassador to the UN François Delattre that

> the Security Council has heard the urgent request made by the Malian President and the Heads of State of West African States, whose soldiers are risking their lives as peacekeepers in Mali. MINUSMA is addressing the threat of terrorist groups. This is an exceptional environment for a United Nations peacekeeping operation, and we are aware of that.[4]

Despite the unanimous vote in favor of Resolution 2295, and the support of Western countries as a whole, some dissenting voices questioned the validity of this mandate adjustment. Russia in particular in particular said it had voted in favor of the resolution in order to "preserve consensus", but accused some delegations of indulging in speculation, by including "ambiguous and vague" wording in the mission's mandate. They were particularly concerned that notions such as "asymmetrical threats" could trigger the use of force and entail risks of damaging the reputation of Blue Helmets.

Some traditional troop-contributing countries, such as Uruguay, also expressed concerns that UN peacekeeping operations are not suited for

[3] Security Council, The situation in Mali S/PV.7727, June 29, 2016.
[4] Ibid.

military counterterrorism operations, insisting that the mission's more proactive mandate should not result in pre-emptive attacks that could alter the nature of its peace operations.[5] Indeed, while Council votes on Mali were unanimous at the time, this should not occult the differences of opinion among Security Council members on whether or not the mission should be involved in military counterterrorism operations, and more generally whether a peacekeeping operation is the right tool for this kind of "asymmetric threat" environment.

In November 2016, the Security Council under the monthly presidency of Senegal, held an unprecedented open debate at ministerial level, in the presence of UN Deputy Secretary-General Jan Eliasson, on the asymmetric and terrorist threats facing United Nations peace operations. MINUSMA, the deadliest of the peacekeeping operations, was on everyone mind. In his introductory remarks, the Senegalese foreign minister argued in favor of more robust mandates, considering that Council Resolution 2295 (2016) strengthening MINUSMA was a step in the right direction, even though Mali noted that its provisions were "slow to materialize", adding that "for greater impact, we need to strengthen our missions in terms of equipment and operational capabilities, and increasingly consider the use of modern technologies."[6] But at the same time, Senegal was slow to deploy its own helicopters promised to MINUSMA for Central Mali, and ultimately canceled their deployment (citing the lack of protective hangars and then the crash of one of its helicopters in MINUSCA in September 2019) and only sent two infantry companies that were based in Mopti.

At the debate, the Minister of Foreign Affairs of the Democratic Republic of the Congo (DRC), Raymond Tshibanda N'tungamulongo, representing the voice of host states, cited MONUSCO's force intervention brigade as an example, which "inflicted a heavy defeat on the former M23 movement, consolidated security in the east of the country and re-established the Mission's credibility with the local civilian population."[7] The Malian government had already called for the creation of a similar intervention brigade to combat terrorist groups in northern Mali, in

[5] Ibid.

[6] Security Council, "Security Council reflects on ways for UN peace operations to address asymmetric and terrorist threats: press releases," CS/12577, November 7, 2016. https://www.un.org/press/fr/2016/cs12577.doc.htm.

[7] Ibid.

addition to the existing French military counterterrorist force already in the country.

Conversely, many other member states taking part in the Council's ministerial debate that day echoed the findings of the 2015 High-Level Independent Panel on Peace Operations (HIPPO) which concluded that, "owing to their composition and character, [UN peace operations] are not suited to engage in military counter terrorism operations." The Chinese representative, while recognizing the call for peace operations to be more active in the fight against terrorism, clearly expressed his preference for strengthening the counterterrorist capabilities of host countries instead. Both China and Russia emphasized the primacy of the basic principles of peacekeeping, and the Brazilian representative warned that "abandoning these principles, in particular the ability to act as an impartial mediator, in favor of an emphasis on military aspects, carries risks for the acceptance of these missions and the safety of Blue Helmets."[8]

This request by the Security Council for MINUSMA to "adopt a more proactive and robust approach" also highlighted a certain gap between decision-makers in New York and the mission on the ground. Aware of the complexity of the asymmetric context, the mission did not wish to extend its military prerogatives beyond its initial mandate. MINUSMA leadership at the time was wary that it would create new expectations on the part of the host state and populations, as well as from some international partners, which the mission would not be able to meet. In an interview with *Le Monde* at the end of 2017, MINUSMA Head of Mission Mahamat Saleh Annadif lamented this gap between distant perceptions and the realities on the ground:

> What I see is that I'm losing men almost every day. But, basically, I see that these critics ignore certain concepts. The real content of a peacekeeping operation covers a global, multidimensional vision that takes into account political progress, good governance, the protection of civilians, the defense of human rights. In other words, it's not limited to the security aspect, which is the only aspect people see. In Mali our peacekeeping mission has a mandate adapted to a given political situation that has evolved. It is deployed in an environment polluted by terrorists, who are imposing an asymmetrical war on it. But the fight against terrorism is not part of MINUSMA's man-

[8] Ibid.

date whereas we are suffering the consequences. We are not a peace-enforcement mission. But would that really be the solution?[9]

In reality, MINUSMA remained very timid in its use of force, despite this new mandate. This was mainly due to the particularity of the asymmetric and terrorist threat it faced—which claimed the lives of some 310 peacekeepers and injured hundreds more between 2013 and 2023, notably but not only due to numerous improvised explosive device (IED) attacks on convoys.

Apart from cases of strict self-defense during attacks on MINUSMA convoys and camps, peacekeepers only really used force once in January 2015. As a group of peacekeepers was caught in the crossfire between two rival rebel groups, one of MINUSMA's Dutch Apache helicopters intervened to destroy the rocket launcher that was firing in the direction of UN troops.[10] In October 2021, British peacekeepers also killed two armed men who had opened fire on their vehicles.

Although MINUSMA never directly confronted armed terrorist groups (it had neither the mandate nor the means to do so), the peace operation nevertheless demonstrated a certain ability to adapt to this environment. In particular, its contingents were trained by the United Nations Mine Action Service (UNMAS) on the risks of IEDs, they were equipped with mine-resistant vehicles provided by donors, and bunkers were built in the camps as shelters during rocket or mortar attacks. Some of the most exposed MINUSMA camps were equipped with *Thalès* Ground Alerter (GA-10) radars and other high-end sensors to detect mortar fire. The tethered observation balloon (A-NSE 350 system) deployed over the Kidal camp was less successful and was eventually withdrawn after it was deemed unsuitable for the weather conditions and replaced by cameras.[11] In 2018, this extensive security system was extended to the UN's "super camp" in Gao, adjacent to that of the French *Barkhane* force.

At the same time, however, attacks on camps and logistical convoys were becoming more complex as terrorist groups grew more sophisticated. Less equipped and trained, African peacekeepers suffered the

[9] *Le Monde*, Mali: "La lutte antiterroriste n'est pas dans le mandat de la Minusma", November 17, 2017.

[10] MINUSMA, "MINUSMA constrained to use force today in Tabankort." January 21, 2015.

[11] François Rihouay, "Au dessus du camp onusien de Kidal, un ballon captif à rude épreuve", Ouest France, February 27, 2016.

greatest loss of life. To meet the challenge posed by the risk of terrorist attacks, MINUSMA had to devote most of its operational capacity to convoy escort and self-protection and moved in the direction of "bunkerization." As a result, the mission's mobility, and hence its ability to implement its mandate, were drastically limited, particularly for civilian components.

The MINUSMA force commander between 2015 and 2017, Danish General Lollesgaard, whose own convoy once came under attack near Timbuktu, repeatedly stressed that 80 percent of his UN force was dedicated to protecting camps and convoys, and that there were very few blue helmets left to implement the civilian protection mandate. In spring 2021, the UN Secretary-General made the following assessment:

> MINUSMA made progress with the installation of overhead protection on its accommodations, the construction of bunkers and the construction of sidewall protection for offices and accommodations. Security enhancement projects, including boom gates, observation posts, ditches and berms, and fencing are in progress … Given the rising number of asymmetric attacks against MINUSMA peacekeepers, Malian Defence and Security Forces and the G5 Sahel Joint Force, the Mission continued to improve its medical evacuation procedures … Since the inception of the Mission in 2013, a total of 140 MINUSMA personnel have died as a result of malicious acts, including 6 during the reporting period. There have yet to be any convictions for the killings of MINUSMA peacekeepers. As of 3 March, four individuals were being held in relation to attacks against MINUSMA. A total of 42 investigative measures are underway and MINUSMA has provided support to the Malian authorities in their efforts to investigate and prosecute 13 cases.[12]

MINUSMA peacekeepers, thanks in part to the implementation of the *Action Plan* following the 2017 Cruz report on "Improving the security of UN peacekeepers,"[13] progressively gained a better ability to respond to attacks. The mission also started communicating more about attacks it countered (even if these still resulted in deaths among UN ranks). A MINUSMA April 2, 2021 press release on the attack on its Aguelhok

[12] Report of the UN Secretary-General on the situation in Mali S/2021/299, March 26, 2021, Para 84–88.

[13] Santos Cruz Report, "Improving Security of United Nations Peacekeepers: We need to change the way we are doing business," December 9, 2017.

camp, emphasized for instance "the courage and bravery of Blue Helmets" and the fact that the "assailants suffered heavy losses."[14]

MINUSMA also faced an issue of perceptions. The presence of Western contingents equipped with modern means rarely seen in UN missions before (in terms of air assets, intelligence gathering and analysis capability, and drones) fueled the perception that MINUSMA could effectively contribute to the fight against terrorism. Several Western troop contributors refused to paint their planes, helicopters and vehicles white, or to wear UN blue helmets/berets and patches, in order to retain military camouflage (and be less easily discernible from the French counterterrorist force). Some of these Western soldiers made no secret of the fact that they would have preferred to join the fight against terrorism alongside the French force, even though their politicians and diplomats had decided otherwise.

"Parallel" Forces or a *Sacred Union* Against Terrorism with *Barkhane* and G5 Sahel

The nature of the relationship between MINUSMA and the French "parallel (counterterrorism) force" *Serval*, then *Barkhane*, was an issue from the outset in 2013 until the departure of the French forces in the summer of 2022. Throughout the years, there were cyclical rapprochements and distancing based to instructions from New York, but also on the different personalities of the UN operation's civilian and military leadership. Although the mandates and resources of the two forces were quite different, there was inevitably a mixing of genres, especially as France tried at times to influence the conduct of UN operations, even though it denies it.

The French force's in extremis support for MINUSMA was authorized by the UN Security Council[15] and was critical, particularly in the build-up phase of the UN operation. On several occasions, French fighter jets were

[14] MINUSMA, "Press Release - Blue Helmets Repel Complex Attack on MINUSMA Camp in Aguelhok," April 2, 2021, https://minusma.unmissions.org/les-casques-bleus-repoussent-une-attaque-complexe-contre-le-camp-de-la-minusma-à-aguelhok-1.

[15] Security Council Resolution 2100 (2013) of April 25, 2013, which decided to create MINUSMA as of July 1, 2013, also authorized "the French army, within the limits of its capabilities and in its areas of deployment, to use all necessary means, from the start of MINUSMA's activities until the end of the mandate authorized by this resolution, to intervene in support of elements of the Mission in the event of grave and imminent danger at the request of the Secretary-General, requests France to report to it on the implementation in Mali of the present mandate." Para 18. First, the *Serval* force, which in July 2014 will become a regional *Barkhane* force of over 3000 troops, including 1000 in Mali, as well as the special forces of Operation *Sabre*.

sent to demonstrate their dissuasive force during complex attacks on MINUSMA camps, notably in the Kidal region. But this "parallel" presence also created major perception challenges. The UN Secretary-General of the UN noted this in advance of MINUSMA's deployment:

> it is critical that a clear distinction be maintained between the core peacekeeping tasks of an envisaged United Nations stabilization mission and the peace enforcement and counter-terrorism activities of the parallel force that will necessarily need to be established to preserve the hard-won security gains achieved so far. Any blurring of this distinction would place severe constraints on the ability of United Nations humanitarian, development and human rights personnel to safely do their work. If this were to happen, the United Nations would find it difficult to mount the kind of comprehensive system-wide response required to address the political, social and economic root causes of the multifaceted crisis in Mali.[16]

In practice, many of MINUSMA's African contingents representing approximately 6000 troops inherited from AFISMA had and continued to benefit from essential operational support from French forces in the early months of the UN mission. On the ground, *Serval* embedded some operational support and liaison detachments (*Détachements de Liaison et d'Appui Opérationnel*—DLAO) with some of MINUSMA's African contingents.[17] The French counterterrorist force was also sharing camps (and some logistics supplies) with UN peacekeepers, including in Kidal, Timbuktu, Tessalit, and Ménaka.[18]

In 2013, so-called joint military operations were organized between *Serval*, the Malian army and MINUSMA.[19] UN headquarters later requested that the military operations of the different forces be called "coordinated" rather than "joint", to avoid MINUSMA being assimilated to a counterterrorist force. In practice, however, the distinction was not always clear-cut, as both the French army and the UN stabilization

[16] Report of the Secretary-General on the situation in Mali S/2013/189, March 26, 2013, paragraph 100.

[17] Each DLAO is made up of 32 soldiers, including a command team, engineers, infantry, a full medical team, mechanics and air controllers to request air support if necessary. Ministry of Defense, "Mali DLA and DAO merge and become DLAO", DICOD, February 13, 2014.

[18] Alexandra Novosseloff & Lisa Sharland, "Partners and Competitors: Forces Operating in Parallel to UN Peace Operations", International Peace Institute, November 2019.

[19] Le Monde, "A large-scale military operation underway in Mali," October 2013.

mission were supporting the Malian state and its defense and security forces on the ground.

In Bamako, France seconded twenty military staff officers within MINUSMA, officially to help the operation get off the ground and to ensure proper coordination with *Serval.* France later retained a number of key positions within MINUSMA force headquarters—including the force Chief of Staff and the U3-5 Chief[20]—as well as at sector level. This allowed for better information sharing (also ensured more formally by a *Serval* liaison officer) and better coordination with the non-UN "parallel force." But above all, it enabled France to influence the planning and conduct of UN operations, at least on the margins, so as to avoid MINUSMA "getting in the way" of French counterterrorist operations, but also, when the opportunity arose, to ensure that the presence of peacekeepers could be used to good effect in such operations, by blocking the possible flight of so-called terrorist groups, for example.

Some French generals who held the position of MINUSMA force Chief of Staff even organized weekly meetings of all French military staff officers seconded to MINUSMA, even if it fed suspicions that France was operating a parallel chain of command and seeking to "use" MINUSMA to serve French interests. The fact is that the presence of these French officers—and Western staff officers in general—largely contributed to making MINUSMA more effective in the planning of military operations, but also at times to better coordinate its actions with those of *Barkhane* and Malian forces, sometimes pushing the mission to the limits between peacekeeping and counterterrorism.

Some of MINUSMA's European contingents, including the Dutch *Task Force* deployed as early as 2013 and comprising 450 soldiers, including special forces and *Apache* and *Chinook* helicopters (based in a Dutch camp adjacent to that of the *Serval* operation in Gao), also seemed more interested in collaborating with French forces than with the UN mission's non-European contingents. Later, some European countries chose to contribute to both *Barkhane* and MINUSMA (Denmark and United Kingdom whose *Chinook* helicopters deployed with *Barkhane* were later used 50 percent of the time to support the 250 British peacekeepers of the reconnaissance unit deployed as part of MINUSMA[21]), and/or offered

[20] Planning and conducting peacekeeping operations.

[21] Laurent Lagneau, "Barkhane/Mali Les Hélicoptères De Transport Lourd Danois Entament Leur Mission Sur Les Chapeaux De Roues - Zone Militaire", *Zone Militaire*, January 15, 2020.

logistical and aeromedical support to *Barkhane* while being part of MINUSMA (The Netherlands and Canada notably). The British soldiers who joined MINUSMA in 2020 also seemed to have adopted a more proactive stance toward terrorist groups in general, unlike most other peacekeepers who felt that it was not their mandate.[22]

Information sharing between MINUSMA (notably the ASIFU unit, made up of eighty European intelligence officers) and the French counter-terrorist operation also raises questions. John Karlsrud suggests that ASIFU may have developed and shared "targeting packs" of groups and individuals considered a threat to the mission. The latter could then become the target of French counterterrorist operations.[23] An internal UN study on lessons learned from ASIFU in 2015 warned, "given that information sharing with Operation *Barkhane* may have political implications, it would seem that decisions on whether or not to share information should be taken at the political level, that is, by the mission's senior leadership and informed by UN policy, rather than by ASIFU commanders."[24]

Resolution 2391 (2017) further reinforces this ambiguity, by adding to the UN mission's mandate "operational and logistical support for MINUSMA" in favor of the joint G5 Sahel (FC-G5S) counterterrorist force long requested by France.[25] The *Serval* and *Barkhane* forces have been supporting "joint cross-border military operations" by the military forces of the five countries in the sub-region (Burkina Faso, Mali, Mauritania, Niger and Chad). This later became the G5 Sahel joint force (FC-G5S) in November 2015, following the announcement by the G5 Sahel heads of state at the N'Djamena summit.[26]

The Security Council in its Resolution 2359 (2017) welcomes "the deployment of the FC-G5S throughout the territories of its contributing

[22] Hazel Shearing, UK troops seize IS weapons in Mali operation, May 2021, BBC News.

[23] John Karlsrud, "Towards UN Counter-Terrorism Operations?", *Third World Quarterly* 38, n° 6 (June 3, 2017): 1215–31, https://doi.org/10.1080/01436597.2016.1268907.

[24] UN internal study on lessons learned from ASIFU in 2015, consulted in 2019.

[25] This support "would be granted to the defense and security forces of the G5 Sahel states only when they are operating on Malian territory as part of the Joint Force", including "primary and secondary medical evacuations, as well as access to essential consumables (fuel, water and rations), the use of UN lifting and earthmoving equipment and engineering equipment, and MINUSMA uniformed engineering units to help prepare the operational bases of the Joint Force in Mali." Security Council Resolution 2391, S/RES/2391 (2017), December 8, 2017. Paragraphs 12 & 13.

[26] Cyril Robinet, "Genèse de la force conjointe du G5 Sahel", IFRI Blog Afrique, January 16, 2018.

countries, with up to 5000 military and police personnel, with a view to restoring peace and security in the Sahel region"[27] but the fact that MINUSMA provides it with direct support, commits the peacekeepers more directly on the ground. The FC-G5S headquarters in Sévaré will moreover be targeted by a suicide attack on June 29, 2018 and will subsequently be moved to Bamako.[28]

This support to the G5 Sahel joint force has also given rise to open opposition between France and the United States at the United Nations. The Trump administration (backed by the UK) even threatened to veto a draft resolution drafted by France supporting the creation of this force, as it feared that this would imply UN assessed funding and thus an additional financial strain on the peacekeeping budget, more than a quarter of which is covered by the United States.[29]

According to the United States, a Security Council resolution authorizing this force to "use all necessary means" under Chapter VII of the Charter to "combat terrorism, drug trafficking and human smuggling" was not required, as troops from Burkina Faso, Chad, Mali, Mauritania and Niger already had the legal authority to carry out missions on their respective national territories. Furthermore, the mandate of the force set out in the French draft resolution would be too broad and vague, with no supervisory mechanism, and would lack an exit strategy. It is for this reason that Resolution 2391 (2017) includes specific mention of the fact that "the United Nations should be fully reimbursed through a funding mechanism coordinated by the European Union for the purpose of coordinating international voluntary contributions in support of the Joint Force."[30]

France's failure to obtain authorization for UN financial support for the G5 Sahel Joint Force prompted then French Foreign Minister Jean-Yves Le Drian to raise the question of the UN's role in the fight against terrorism during an annual debate on the reform of peacekeeping operations:

[27] Security Council Resolution 2359, S/RES/2359 (2017), June 21, 2017.

[28] Le Monde, "Mali: cinq morts dans une attaque contre le QG de la force militaire G5 Sahel," June 29, 2018, https://www.lemonde.fr/afrique/article/2018/06/29/mali-au-moins-six-morts-dans-une-attaque-contre-le-qg-de-la-force-militaire-g5-sahel_5323417_3212.html.

[29] Colum Lynch, "Trump Weighs Vetoing France's African Counterterrorism Plan", Foreign Policy, June 13, 2017.

[30] Ibid Para 13 c).

> We need to find the right tools to deal with evolving threats and complex crises ... In Mali today, we are faced with a paradox. On the one hand, we have MINUSMA, an operation under Chapter VII. On the other hand, we have the G5 Sahel joint force, which has a counter-terrorism mandate but little or no resources and has to plead for ad hoc support from its partners, with no guarantees and no predictability. This situation is unsatisfactory. And if giving MINUSMA a counter-terrorism mandate is not the solution, then we need to work on a different architecture.[31]

Implementing its mandate to "contribute to the re-establishment of State authority throughout the country"[32] often puts the UN mission in the difficult position of supporting the Malian army (first under the Malian flag, then also under the mandate of the G5 Sahel sub-regional force from 2017 onwards), all the more so when serious human rights violations are committed during counterterrorism operations.[33]

This is why the UN set up a *Human Rights and International Humanitarian Law Compliance Framework* to guide its support for the G5 Sahel joint force, over and above the due diligence policy on human rights that already applies in any context where the UN provides support to non-UN security forces (the Human Rights Due Diligence Policy or *HRDDP*).[34]

With the growing number of military forces operating in Mali, liaison officers were exchanged between forces, and the forces started meeting on a quarterly basis in the *Instance de Coordination Militaire du Mali (ICMM)* between MINUSMA, Operation *Barkhane,* the EU training mission (EUTM), the G5 Sahel joint force and the Malian army, starting in 2018 at the level of force commanders,[35] "with a view to addressing

[31] Security Council, United Nations peacekeeping operations: reform of the United Nations peacekeeping system: implementation and follow-up, S/PV.8051, September 20, 2017.

[32] Security Council Resolution 2100, S/RES/2100 (2013), April 25, 2013, paragraph 16.

[33] Matteo Maillard, "Les forces de sécurité maliennes accusées de 101 exécutions extrajudiciaires par l'ONU," Le Monde, May 2, 2020.

[34] Office of the United Nations High Commissioner for Human Rights, Progress Report: OHCHR Support Project to the G5 Sahel Joint Force in the Implementation of the Human Rights and International Humanitarian Law Compliance Framework May 1, 2018–March 31, 2020, August 5, 2020.

[35] MINUSMA, "Quarterly meeting of the Instance de Coordination Militaire du Mali (ICMM)", March 15, 2021, https://minusma.unmissions.org/réunion-des-commandants-des-forces-internationales-au-mali.

however, mission members can lose this protection if they directly take part in hostilities. While there is no doubt that attacks on MINUSMA civilian personnel and attacks on camps housing such personnel constitute war crimes, attacks directed exclusively against MINUSMA military personnel present a greater legal complexity, as the International Commission of Inquiry and the International Committee of the Red Cross (ICRC) have repeatedly emphasized.[42]

If peacekeepers were to become parties to a conflict, not only would they lose their protective status and become legitimate military targets under IHL, but this could also call into question the "integrated" nature of the mission—notably of development and humanitarian components of the UN family whose actions are meant to be coordinated by the MINUSMA deputy head of mission who also wears the hats of Resident Coordinator and Humanitarian Coordinator (DSRSG/RC/HC). Some UN agencies, funds and programs actually sought to distance themselves from the blue helmets and left the MINUSMA camps in an attempt to preserve their independence and humanitarian space.

A certain ambiguity was certainly inevitable in such a complex situation. Nevertheless, we can see that France in the Security Council, instead of following the recommendations of the UN Secretariat, contributed to further reinforcing this ambiguity through Resolution 2295 (2016) by asking MINUSMA to adopt a "more proactive and robust approach" and Resolution 2391 (2017) by asking it to provide operational and logistical support to the G5 Sahel joint force. In so doing, it is also exacerbating tensions within the P3, with the United States and the UK. The late addition of a sanctions regime in 2017, and the way it was managed, would time exacerbate tensions with Russia in particular.

The 2017 Addition of a Sanctions Regime at the Request of the Malian Government

During the second half of 2017, the "international community" stepped up its efforts to encourage the Malian parties to implement the peace agreement. The *Comité de Suivi de l'Accord (CSA)*, notably appointed the

[42] United Nations Security Council, *Report of the International Commission of Inquiry for Mali* January 29, 2021, paragraphs 825–830. The Commission ultimately "considers this to be a matter of great legal and strategic importance that could be taken up by the Secretary-General."

Carter Center as an independent observer to assess the state of implementation of the agreement. An independent International Commission of Inquiry for Mali was also set up. And above all, the Security Council decided, through Resolution 2374 dated September 5, 2017 (adopted unanimously) to establish a sanctions regime for Mali.

This sanctions regime, although independent of the peace operation on the ground, was designed to complement the action of MINUSMA. The *Group of Experts* which supported the work of the Council's Sanctions Committee was also largely dependent on MINUSMA for access (logistics) and for its security. This security had moreover been reinforced after the assassination of two experts from the group working for the UN Sanctions Committee on the Democratic Republic of Congo (DRC) in March 2017.[43]

The sanctions regime for Mali specifically targeted actors who would derail the peace process in Mali, notably by imposing a travel ban and freezing the assets of individuals and entities designated.[44] It was therefore extremely broad and came in addition to other sanctions regime already in existence targeting armed groups already designated as terrorists by the Security Council. These include MUJAO, AQIM, *Ansar Eddine* and its leader Iyad Ag Ghali, and *Al-Mourabitoun*, which were already included on the ISIL (Da'esh) and *Al-Qaeda* Sanctions List established in accordance with resolutions 1267 (1999), 1989 (2011) and 2253 (2015). In

[43] France Inter, "En RDC, deux experts des Nations-Unies sauvagement assassinés et le monde regarde ailleurs," December 2, 2018, https://www.franceinter.fr/emissions/profession-reporter/profession-reporter-02-decembre-2018.

[44] for the following reasons: "a) Taking part in hostilities in violation of the Agreement; b) Taking actions which impede or threaten the implementation of the Agreement, including persistent delays; c) Acting on behalf of, or in the name of, or on the instructions of, a person or entity engaged in the activities referred to in paragraphs a) and b) above, or providing any other form of support or financing to such person or entity, in particular by using the proceeds of organized crime, including the illicit production of and trade in narcotics and their precursors originating from Mali or in transit through the country, trafficking in human beings, migrant smuggling, arms smuggling and trafficking in cultural goods; d) Preparing, ordering, financing or carrying out attacks... e) Obstructing the delivery of humanitarian aid destined for Mali, access to such aid or its distribution in the country; f) Preparing, ordering or committing in Mali acts contrary to international human rights law or international humanitarian law, or which constitute human rights violations or abuses... (g) The use or recruitment of children by armed groups or armed forces in violation of international law, in the context of the armed conflict in Mali." Security Council, Resolution 2374, S/RES/2374 (2017), September 5, 2017.

February 2020, the Security Council will add Amadou Kouffa, founder of the *Macina Katiba*, part of the JNIM, to this list.[45]

The idea of a sanctions regime for Mali was not new. Some member states had not hesitated to brandish the threat of sanctions when the CMA proved reluctant to initial the Algiers agreement, with the Malian government also calling on the "international community" to "isolate [...] by imposing sanctions" on "radical and extremist individuals" who would not sign the agreement.[46] Already in 2015, the Security Council in Resolution 2227 (2015) declared its

> readiness to consider targeted sanctions against those who take actions to obstruct or threaten the implementation of the Agreement, those who resume hostilities and violate the ceasefire, as well as those who attack and take actions to threaten MINUSMA.[47]

However, it took a formal request in the form of a letter from the government of Mali in August of the same year for the Security Council to finally decide to set up this sanctions regime and a *Group of Experts* to support the work of the Council's Sanctions Committee. This however gave the government the impression that it would be above this regime, whereas theoretically, as a party to the conflict and signatory to the peace agreement, the Malian government could also be exposed to sanctions if it were responsible for "derailing the peace process in Mali."

The Security Council is divided on the issue of sanctions. Russia, itself a target of international sanctions, is opposed to sanctions in principle. However, it initially voted in favor of the sanctions regime, but expressed concern about the usefulness of sanctions. Russia's deputy ambassador Petr Iliichev made it abundantly clear:

> We are always against sanctions regimes, especially in this particular situation, when one of the parties to the peace and reconciliation agreement asks

[45] United Nations, "ISIL (Da'esh) and Al-Qaeda Sanctions Committee adds entry to its sanctions list - press releases." SC/14097, February 4, 2020, https://www.un.org/press/fr/2020/sc14097.doc.htm.

[46] Mali's Minister of Foreign Affairs Abdoulaye Diop, in his speech to the UN Security Council on April 9, 2015, S/PV.7425.

[47] Security Council, Resolution 2227, S/RES/2227 (2015), June 29, 2015, Para 3.

for sanctions against the other two parties […] When one party asks for another (to be sanctioned), we can expect this agreement to collapse.[48]

In practice, despite the Malian government's many delaying tactics and *fait accompli* policy, the Security Council's Sanctions Committee, itself made up of member states, naturally focused its attention on non-state actors (armed groups) rather than personalities close to the government. This contributed to the Committee being perceived as biased and an instrument of France, the only country who effectively suggested names of individuals to be sanctioned.

No person or entity was designated during the first year of the new sanction regime by the committee initially chaired by Sweden. Creating a sanctions regime without designation can thus play a dissuasive role that seems to have worked initially, as armed groups generally behaved better.[49] However, to be considered credible by the parties to the conflict, a sanctions regime must necessarily lead to designations generally proposed by member states. Conversely, designations have relatively little impact beyond the symbolic in Mali when it comes to non-state actors, since they have few assets (and even fewer abroad) and rarely travel by any means other than road, apart from the leaders of armed groups who have signed the peace agreement.

France proposed the first three individuals to be listed under this sanction regime on December 20, 2018. These were three Malians affiliated with three different armed groups signatories to the peace agreement, accused respectively of having established illegal checkpoints in the Timbuktu region and of being involved in JNIM terrorist attacks against the Malian army (for the CPA), of being involved in drug convoys in northern Mali and ceasefire violations in Kidal (for the GATIA, close to the Malian government), and of involvement in terrorist activities and migrant trafficking (including sexual exploitation of migrants) for a businessman (affiliated to the HCUA, member of the CMA)[50] whose house

[48] RFI, "France lobbies for UN Security Council debate on Mali sanctions", September 2, 2017.

[49] Kristen Boon, "For More Effective Sanctions, Time to Examine Question of Termination," April 24, 2014.

[50] They are Mohamed Ousmane Ag Mohamedoune, Secretary-General of the Coalition du peuple pour l'Azawad (CPA), Ahmoudou Ag Asriw, an executive of the Groupe d'autodéfense touareg imghad et alliés (GATIA), and Mahamadou Ag Rhissa, a businessman belonging to the Haut conseil pour l'unicité de l'Azawad (HCUA). See Morgane Le Cam, "Au Malides signataires de l'accord de paix accusés de terrorisme", Le Monde, August 28, 2018, Jeune Afrique, "Trois Maliens" contrevenant à l'accord de paix "frappés par des sanctions de l'ONU", December 21, 2018.

had already been raided by the *Barkhane* force a year earlier.[51] Following these initial appointments, the Security Council noted with satisfaction in June 2019 that:

> More progress has been made in the last nine months than in the first few years following the signing of the Agreement, a period marked by slow implementation [and] that a certain degree of political will combined with international pressure, notably the prospect of sanctions, played an important role in achieving these positive results.[52]

Researcher Bertrand Ollivier explains, "by sanctioning low-profile personalities, the Committee exerted pressure on the senior public officials of the signatory parties." The sanctions regime is even said to have helped unlock the application of certain measures in the Agreement, notably the implementation of the first stage of the DDR program at the end of 2018,[53] and the fact that the CMA loosened its control over the management of Kidal and began to transfer powers to the interim authorities that resulted from the peace agreement.

In July 2019, five more people were added to this sanctions list, but this time more important and influential members of the signatory armed groups, namely MP Mohamed Ould Mataly—and member of the *Platform* armed groups; Mahri Sidi Amar Ben Daha—alias Yoro Ould Daha—military leader of the *Platform*, which he represented within the operational command mechanism (MOC) before being assassinated a few months later;[54] Mohamed Ben Ahmed Mahri—alias Rougi; Houka Houka Ag Alhousseini—former head of the Islamic police in Timbuktu during the occupation of this city by groups linked to Al-Qaeda in 2012, tried in 2020 by the ICC;[55] and Ahmed Ag Albachar.

This second series of designations provoked stronger reactions from armed groups and certain local communities, even contributing to a

[51] Sahelien.com, "Kidal: assaut contre les domiciles d'un riche "homme d'affaires", sept personnes arrêtées", October 1, 2017.

[52] United Nations, "Mali In the face of delays in implementing the Peace Agreement, the Security Council extends sanctions and mandate of Group of Experts-press release." CS/13933, August 29, 2019.

[53] Bertrand Ollivier, "Note d'analyse: Le comité des sanctions des Nations unies pour le Malia double-edged sword", ThucyDoc n° 19, March 6, 2019.

[54] RFI, "Mali pro-government military leader Yoro Ould Daha killed", February 10, 2020.

[55] Jeune Afrique, "Mali Quand l'ancien juge islamique de Tombouctou au temps de l'occupation jihadiste se pose en homme de paix", September 13, 2019.

certain mistrust of MINUSMA. In particular, the CMA, the *Platform* and the Lehmar Arab community voiced their dissatisfaction and incomprehension at the discrepancy between the accusations made against these individuals and local realities.[56] Demonstrations denouncing the injustice of the sanctions against Mohamed Ould Mataly were even organized in Gao. Mohamed Ould Mataly, a member of parliament belonging to President Ibrahim Boubacar Keïta's party and accused of instigating demonstrations in Gao against the installation of interim authorities in the region, as provided for in the peace agreement, is the best-known of these individuals. He was a regular visitor to MINUSMA offices and traveled on its planes between Bamako and Gao. As a result, he could no longer board MINUSMA aircraft or commercial flights out of Mali.

However, this did not prevent Mohamed Ould Matali from being appointed in March 2021 as advisor to Colonel Malick Diaw, President of the National Transitional Council (CNT), the legislative body of Mali's transitional government, in defiance of the UN Security Council.[57] The Mali sanctions regime lasted until August 2023, when Russia vetoed its renewal (thirteen Security Council members voted in favor of the resolution, while China abstained) in the wake of the Malian decision to end MINUSMA. During the final year of the sanctions regime, the *Groups of Expert* monitoring its implementation had reported "grave human rights abuses" by the Malian troops and its foreign security partners, believed to be Russia's *Wagner Group*. Deputy US Ambassador to the UN Robert Wood told the Council that Russia wanted to eliminate the independent monitoring "to stifle publication of uncomfortable truths about Wagner's actions in Mali, which require attention" while Russia responded that it was "upholding the interests of the affected country—Mali, as the council is supposed to do."[58]

[56] Liza Young & Michel Luntumbue, "Sanctions ciblées et maintien de la paix: quelle contribution à l'exécution du mandat de la MINUSMA?", Observatoire Boutros-Ghali Observatory on Peacekeeping, June 2020.

[57] RFI, "Mali: polémique après la nomination de Mohamed Ould Matali comme conseiller du président du CNT", March 18, 2021.

[58] Reuters, "UN sanctions in Mali to end after Russia vetoes resolution," 31 August 2023.

The Controversial 2018 Independent Strategic Review Divides France and the US

In June 2017, the Security Council extended MINUSMA's mandate for another year without any significant changes this time. However, some of its members expressed concern at the lack of progress in implementing the Peace Agreement and the deteriorating security situation, suggesting that it was time to reflect on the mission's future. The Council therefore requested the Secretary-General to:

> develop within 180 days after the adoption of this resolution a mission-wide strategic plan that: (i) articulates a concrete phased approach to implementation of MINUSMA's mandate; (ii) presents a transition plan with a view to handing over relevant tasks to the UNCT based on their mandates and comparative advantages as well as a mapping of capabilities, and also with a view to a possible long-term exit strategy of the Mission on the basis of improved security and political conditions as well as the progress in the implementation of the Agreement.[59]

The Secretary-General therefore requested that an independent strategic review of MINUSMA be undertaken in early 2018, as he had judged that in a context of growing insecurity, it was indeed time to reassess the mission's presence in Mali.[60] The idea of conducting "independent" strategic reviews at key moments in the life of a peace operation is one of the innovations introduced by Secretary-General Guterres on his arrival at the helm of the UN in January 2017.

Initially, these reviews were to be totally independent of the UN Secretariat, providing the organization with an outside view of the strategic direction of its field operations; but following serious concerns expressed by DPKO and certain permanent members of the Council, a compromise was reached whereby the reviews would be led by well-known UN figures (such as former senior mission and/or headquarters managers) supported by a team of Secretariat/DPKO staff, and a *Red Team of* independent experts would provide a more distanced and critical view at various stages of the strategic review team's work (conflict analysis,

[59] Security Council, Resolution 2364, S/RES/2364 (2017), June 29, 2017, Para 48.

[60] United Nations, "Amid growing insecurity, time to reassess UN peacekeeping presence in Mali Security Council told," January 23, 2018.

formulation of recommendations, and articulation of options for mission reconfiguration).

MINUSMA was not the first peace operation to undergo an independent strategic review.[61] Yet this review, led by the former head of the UN mission in South Sudan (UNMISS), Ellen Margarethe Loj, remains to this day the most controversial of all strategic reviews. The reason lies in the stakes and power rivalries exacerbated at the time, between France (penholder of the resolutions on MINUSMA) and the United States in particular; but also to a difference of appreciation between the mission in the field and what was perceived as a review largely driven from UN headquarters in New York with preconceived ideas of what the future of the mission should be.[62] The result was that:

> Despite fierce opposition by some members of the Council (including the US), the [UN] Secretariat decided not to publish the review and only to adopt some of its recommendations. The controversy raised questions about how open-minded or adaptable the UN system could be about political strategy.[63]

France lobbied behind the scenes to prevent the report from being published. The strategic review report has remained confidential until now, and has never been officially shared with member states. The report (which the author has been able to consult) gives an interesting assessment of MINUSMA five years after its deployment, expressing deep concern that the mission has gradually drifted away from its initial peacekeeping objective, now acting as a "catalyst for counter-terrorism operations" and supporting a Malian national army that lacks popular legitimacy and is accused of human rights violations. The report also expresses concerns that this shift is affecting the impartiality of the UN, confusing roles and responsibilities between actors, and endangering peacekeepers, pointing out that with over 500 casualties (including wounded) in their ranks between 2013 and 2018, "the current situation is not tenable."

[61] Daniel Forti, "Learning from recent Independent Reviews of UN Peace Operations", The Global Observatory, April 22, 2021.

[62] This was despite a visit to Mali and meetings with a wide range of stakeholders during a March 14–25, 2018 visit to field bases.

[63] Aditi Gorur, "The Political Strategy of the UN Peacekeeping Mission in Mali" UNU/Stimson Center, 2020.

The report also expresses concerns about the growing gap between an ambiguous mandate calling for a more "robust" and "proactive" posture creating false expectations that MINUSMA should or could conduct counterterrorism operations. While acknowledging that tangible steps have been taken to generate capacity and improve personnel safety and security, the review concludes that MINUSMA "is not in a position to project itself significantly beyond a 20 km perimeter of its camps," thus limiting its activities including those of the mission's civilian components.

The Secretary-General's report of June 2018 took up several of the findings of the independent strategic review:

> As the sole entity with a presence across northern and part of central Mali, and owing to the constraints and delays encountered in carrying out the process envisioned in Security Council resolution 2295 (2016) for the restoration and extension of State authority to the central and northern regions, the assessment of the review team was that MINUSMA had been called upon to undertake a variety of tasks, including having become a logistics and security enabler for non-United Nations security actors. The review team was of the view that the proximity of MINUSMA, and its support role and cooperation with security actors, including counter-terrorism actors, had contributed to the perception that the Mission was engaging in counter-terrorism actions.[64]

However, the recommendations of the independent strategic review were only partially reflected in the Secretary-General's report to the Security Council, which instead emphasizes the need to "refocus on political issues", in particular through the development of a peace pact involving the Government, the Security Council, the United Nations and international partners.

The independent strategic review had gone much further, suggesting that MINUSMA should be reoriented toward clearer political objectives and even reconfigured, proposing three options that involved more or less reducing the mission's presence in northern Mali, underlining that "*business as usual" cannot* continue.[65] The report also suggested a predominantly civilian presence in Central Mali to implement a tailored approach to

[64] Report of the Secretary-General on the situation in Mali S/2018/541, June 6, 2018, paragraph 54.

[65] From 14 bases to 11 (proposing to close Tessalit and Aguelhok in particular, given their little value to the political process, even if France logistical hubs and the presence of Chadian soldiers) at a constant troop ceiling for the first year, with a possible reduction of 30 percent to 7 bases thereafter; or to 5 bases only, while maintaining a rapid reaction capability, should there be no concrete progress in implementing the Agreement.

civilian protection based on dialogue, engagement, community policing and the creation of a protective environment. Finally, it discouraged future co-location with counterterrorism actors including ***Barkhane***.

This controversy gave rise to a meeting of the Security Council, during which Under-Secretary-General for Peacekeeping Operations, Jean-Pierre Lacroix, merely stated that "the Mission is also to assess its current physical footprint against political priorities, factoring in the constraints posed by the operational theatre, the difficulties in generating resources for the United Nations efforts in Mali [...] Any changes in the Mission's configuration will be carefully evaluated, in consultation with partners on the ground, and a timeline will be developed to ensure a sequenced approach."[66]

The French representative Mr. Michon, speaking immediately afterwards, simply said: "Given those many challenges, the renewal of the MINUSMA mandate for another 12 months is absolutely essential to the stability of Mali and the region. The recommendations of the strategic review, which are included in the Secretary-General's report, provide us with an opportunity to ensure that the resources that we are collectively harnessing in Mali are used in the most effective way possible. We want to make the best use of them."[67]

The representative of the United States Ms. Tachco, who obviously had access to the report of the independent strategic review, offers a very different reading, stressing that "the number of casualties, combined with the limited progress on the ground, shows that we need to think seriously about what we expect from MINUSMA" and that the strategic review "has launched a difficult discussion" and that "it is time to act, given the observations that have been made by the Council over time and the useful and specific recommendations made by the team that carried out this review." She thus calls for a reconfiguration of MINUSMA:

> MINUSMA operates in an incredibly challenging security environment, with few resources available to focus on initiatives beyond the protection of the Mission itself. Given that reality, it is time to streamline the mandate to increase support to political engagement and scale down the Mission's focus on less critical activities. Such streamlining will necessitate tough choices as we seek to eliminate, refine or sequence mandated tasks that are either not directly supporting the political process or have stalled altogether. We, like others, are also worried about growing instability in central Mali, and we

[66] Security Council, The situation in Mali S/PV.8288, June 14, 2018.
[67] Ibid.

> support the review's recommendation that some reconfiguration of the force in the north is necessary. The consolidation of bases could reduce troop exposure and free MINUSMA resources to complete tasks beyond force protection. Such an undertaking ought to be conducted at the discretion of the Mission leadership, with the safety of the Mission and its personnel in mind [...] The strategic review also raises the concern that we share about the parameters of service provision to non-United Nations entities. The report (S/2018/541) rightly points out that support to non-United Nations entities risks jeopardizing the Mission's impartiality and its legitimacy in the political space.[68]

France once again managed to secure a reprieve for MINUSMA, despite the offensive by Donald Trump's US administration to reduce the overall budget for UN peace operations. The mission's mandate was extended for another year, until June 30, 2019, with no change to the troop ceiling of 13,289 military personnel and 1920 police officers. There was no major change in direction either, as the Council "decides that MINUSMA's strategic priority remains to support the implementation by the Government, the armed groups the *Platform* and the CMA, as well as by other Malian stakeholders, of the Agreement."[69]

Instead of refocusing on political issues, the Council even specifies what it means by a "proactive and robust approach" already in the mandate since 2016, in paragraph 34:

> Requests MINUSMA, in pursuit of its relevant priority tasks and active defence of its mandate, to continue anticipate and deter threats and to take robust and active steps to counter asymmetric attacks against civilians or United Nations personnel, to ensure prompt and effective responses to threats of violence against civilians and to prevent a return of armed elements to those areas, engaging in direct operations pursuant only to serious and credible threats.[70]

That said, the Council also initiated a pivot toward Central Mali without yet making this a second strategic priority, and "encourages the reconfiguration of MINUSMA posture to optimize and rebalance the uniformed and civilian presences in the Central region, at the discretion of the SRSG,

[68] Ibid.

[69] Security Council Resolution 2423, S/RES/2423 (2018), June 28, 2018. Paragraphs 24–34.

[70] Ibid.

in close consultation with the Force Commander and without impeding its ability to pursue its strategic priority in the North," and adds as one of its "priority tasks" to "support to the restoration of State authority in the Centre" of the country, including to "support the redeployment of the MDSF in the Centre of Mali."[71]

This is the compromise France negotiated with the United States, more focus on protecting civilians and central Mali in exchange for not lowering the mission's troop ceiling and budget. In other words, MINUSMA had to do more, but with the same human and financial resources.

The United States also included in the Resolution the seeds of reflection on a possible future transition (understand downsizing toward a future exit), by requesting the Secretary-General to report every six months in his periodic reports on the development and implementation of the integrated strategic framework that should also include a

> transition plan with a view to handing over relevant tasks to the UNCT based on their mandates and comparative advantages as well as on a mapping of capabilities and gaps, a resource mobilization strategy that involves the full range of multilateral and bilateral partners, as well as a possible long-term exit strategy of the Mission on the basis of improved security and political conditions as well as of progress in the implementation of the Agreement.[72]

US Pushed for a Second Strategic Priority on Central Mali but with No Additional Resources

Resolution 2423 (2018) had already paved the way for a greater focus on the protection of civilians and central Mali in particular. While Resolution 2480 (2019) reiterates that "MINUSMA's main strategic priority remains to support the implementation of the Agreement," it officially endorses a "second strategic priority" which aim is " to protect civilians, reduce inter-communal violence, and re-establish State authority, State presence and basic social services in Central Mali."[73] The text goes on to detail "priority tasks" such as support for stabilization and the re-establishment of state authority in central Mali, and the protection of civilians.

[71] Security Council Resolution 2423, S/RES/2423 (2018), June 28, 2018. Paragraphs 36 and 38.

[72] Security Council Resolution 2423, S/RES/2423 (2018), June 28, 2018. Para 71.

[73] Security Council Resolution 2480, S/RES/2480 (2019), June 28, 2019. Paragraph 20.

The massacre of 160 civilians on March 23, 2019 in the Fulani village of Ogossagou in central Mali by traditional Dozo hunters reinforced the influence of the "third UN" in favor of prioritizing the protection of civilians by MINUSMA. Under pressure from the United States, MINUSMA was asked, in addition to its quarterly report to the Council, to also report to the Security Council every six months in the form of a letter on the mission's performance, and to start thinking about a transition plan for the mission to delegate some of its tasks to the UN country team and the Malian government.[74]

During the Council debate in June 2019, the representative of the United States insisted that "This is no ordinary mandate renewal, because MINUSMA is no ordinary peacekeeping Mission," explaining that the Mission, then numbering over 16,000 uniformed and civilian personnel with a budget of over $1 billion a year, was operating in a difficult, dangerous and asymmetric context. Expressing his disappointment at the lack of progress in implementing the peace agreement, he welcomed the inclusion of new specific and measurable *benchmarks* focusing primarily on the political and security pillars of the agreement, which would be assessed six months later already.[75]

France's Permanent Representative to the United Nations François Delattre, for whom it was the last appearance at a Security Council session before leaving his post, preferred to emphasize that the resolution "outlines a clear view of MINUSMA's future by establishing that its withdrawal strategy is based on the redeployment of the Malian armed forces across the territory of Mali as well as the full operationalization of the Joint Force of the Group of Five for the Sahel (G-5 Sahel), without in any way jeopardizing the concerted efforts that have been made thus far to stabilize Mali and the Sahel." He was supported by the delegate from one

[74] Ibid. Paragraphs 64–65. "i) a letter containing exclusively and in depth: i) information on security problems in Mali mission operations, contingent performance and rotations, as well as an update on discussions at the Coordination Forum in Mali on the coordination of security responsibilities ... ; ii) an update on the implementation of the above mentioned integrated strategic framework, including a transition plan to delegate certain tasks to the UN Country Team, in line with the mandates and relative advantages of each entity and based on an inventory of capacities and gaps, a resource mobilization strategy involving all multilateral and bilateral partners, as well as a possible long-term drawdown plan for the Mission, taking into account improvements in the security and political situation, as well as progress made in implementing the Agreement."

[75] Security Council, The situation in Mali S/PV.8568, June 28, 2019.

of France's traditional allies, Côte d'Ivoire, who expressed concern that exit strategy issues were being discussed prematurely, arguing that the situation in Mali remained serious, and that such decisions should only be considered when the situation on the ground improves.[76]

While France was concerned by the American position, Germany, an elected member of the Security Council for 2019–2020, proposed to share the pen with France on the resolution. France and Germany had just announced symbolic "twin" presidencies for March and April 2019, taking turns in holding the presidency of the Security Council, but working together to advance issues of common interest with coordinated work programs.[77] In 2019, the UK has for the first time symbolically invited Germany, to become co-pen on the AU-UN hybrid operation in Darfur (UNAMID). France, however, rejected Germany's request on Mali, for fear that this would complicate already difficult negotiations with the United States, even if it meant upsetting a major ally that had deployed troops to MINUSMA at its request.

France finally conceded to the United States the *benchmarks* and the start of discussions on a possible future transition, but with no precise timetable. A few months later, however, at the Council meeting of January 15, 2020, the United States suggested that the Council "must recognize that peacekeeping missions are not the answer to growing terrorist threats in Mali" and proposed a new "clear-eyed assessment of MINUSMA to determine how the Mission most effectively complements other security activities in the region" to be carried out by June 2020. The US suggested, "it is therefore time for the Council to begin developing an alternative approach to addressing growing instability in Mali" and that "we should consider deprioritizing MINUSMA's support in implementing the agreement and instead focus the Mission on protecting civilians. That would allow MINUSMA to direct its resources to the more densely populated and strategically important center, where it can have an immediate impact. Additionally, the Mission can reduce its size, allowing Member States to apply resources to more effective efforts in the region."[78]

In practice, on the ground, the mission was already working in Central Mali where it had a small civilian office in Mopti since 2013, despite strong

[76] Ibid.

[77] Permanent Representation of France to the United Nations in New York, March 2019: French Presidency of the Security Council.

[78] Security Council, The situation in Mali S/PV.8703, January 15, 2020.

resistance from the Malian government, which had long denied the deterioration of the situation in this region, close to the capital Bamako. As early as 2016, and following the August 7, 2015 attack on the *Byblos* Hotel in Sévaré in the Mopti region[79] and a strategic review conducted by DPKO (which recommended an additional 2049 troops for the military component),[80] the Mission had anticipated the addition of the second strategic priority, particularly in terms of logistics planning, even if it did not yet have peacekeeping troops to deploy to Central Mali.

The Council thus validated the actions already initiated by MINUSMA. But the problems in central Mali are extremely complex and very different from those in northern Mali, and were not covered by the Algiers peace agreement,[81] making the task all the more difficult for UN employees. In view of the addition of the second strategic priority, but also of the pressure from the Security Council to accelerate the shift to the Center, the *Mission Concept* was revised in late 2019 to reflect this major change. For the first time it was UN Headquarters Integrated Operational Team (IOT) and not the Mission itself that led the process.[82]

The challenge for the Mission was to translate these major strategic orientations into more operational terms at a time when it had neither more troops nor more financial resources. That said, despite the tensions between the United States and France over financial aspects (including funding for the G5 Sahel joint force), the permanent members were able to maintain a degree of political unity behind the two strategic priorities they assigned to MINUSMA. What was lacking, however, was the political commitment of these same permanent members to support the mission in Mali, as the United States and France did in the context of the 2015 Algiers peace negotiations, to encourage the parties to sign the agreement.

[79] France 24, "Mali l'attaque de l'hôtel de Sévaré revendiquée par un cadre jihadiste", August 11, 2015.

[80] An integrated strategic review mission, led by the Department of Peacekeeping Operations, was carried out from March 14 to 20 to assess the appropriateness of MINUSMA's mandate and structure three years after its deployment, in the light of political and security developments, including the signing of the Agreement for Peace and Reconciliation in Mali. Security Council, Report of the Secretary-General on the situation in Mali, S/2016/498, May 31, 2016.

[81] With extremist groups exploiting local inter (farmers and herders) and intra-community conflicts, with a proliferation of armed groups and community militias.

[82] Aditi Gorur "The Political Strategy of the UN Peacekeeping Mission in Mali" UNU/Stimson Center, 2020.

In response to the Security Council's addition of a second strategic priority for central Mali while maintaining the same troop ceiling and overall budget in 2019, MINUSMA developed an *Adaptation Plan* for its Force and civilian components. The creation of a new military sector for the center of the country in June 2019 meant that some of MINUSMA's military capabilities would be relocated to the mission's regional office based in Mopti.[83] But these military capabilities remained very limited for a very large region[84] where road travel is limited by the threat of IEDs and by the rainy season. The MINUSMA force succeeded in concentrating significant resources on the Central sector, albeit on an ad hoc basis, by borrowing certain units from other sectors.[85]

The military *Adaptation Plan* devised by Swedish General Dennis Gyllensporre, MINUSMA's Force Commander from summer 2018 to fall 2021, was intended to be a game-changer, enabling the mission to operate more securely in the face of asymmetric threats by making peacekeepers more mobile and flexible with additional air and surveillance assets (ISR) as well as rapid reaction forces. But given that the troop ceiling had remained the same, the mission had to send back some infantry companies (the worst performers) to make way for new more capable units. And new units were promised at the May 2020 force generation conference (Fig. 3.1).[86]

[83] Whereas military contingents previously depended on either Timbuktu or Gao.

[84] Due to camp protection requirements, only two infantry companies from Senegal based in Mopti and two infantry companies from Togo based in Douentza are actually able to carry out MINUSMA's protection mandate in the Center. Only one company from each of these battalions can set up a temporary operational base (TOB) at a time. The force's ability to react is therefore very limited. Air assets are also lacking, with a unit of light military attack helicopters from Timbuktu-based Salvador stationed part-time in Mopti to support this new sector.

[85] The first major operation of this type was Operation *Folon 1* in February 2019, for which the Timbuktu-based Swedish Intelligence, Surveillance and Reconnaissance (ISR) company was relocated to Mopti for several weeks at great cost in terms of air assets and advanced medical teams. Similarly, during Operation *Mangoose* in June 2020, Egyptian special forces and a Pakistani military helicopter were temporarily redeployed from Sector West (Timbuktu) to the center of the sector to reinforce Senegalese and Togolese infantry companies. The added value of a multidimensional mission such as MINUSMA nevertheless remains its integrated military-civilian dimension, and some of the most recent operations, such as Operation *Buffalo Dry* in early 2021, have enabled civilian components to carry out their work thanks to the deployment of a TOB.

[86] Including a level-2 hospital and Pakistani military helicopters, rapid reaction forces, and unarmed drones promised by China to be based in Timbuktu.

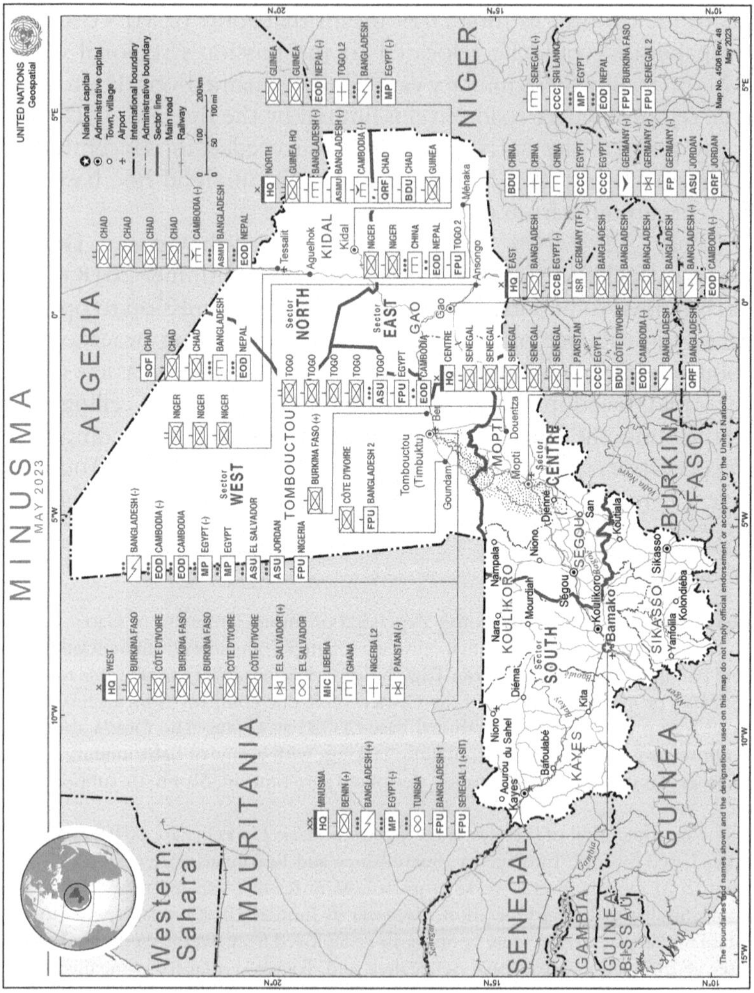

Fig. 3.1 MINUSMA deployment map (May 2023)

The COVID-19 pandemic and the coup d'état in August 2020 however slowed down the implementation of the *Adaptation Plan*.[87] In Resolution 2531 (2020), the Security Council expressed "its full support for the continued implementation of the MINUSMA adaptation plan" and called on member states to "contribute to the plan by providing the capabilities necessary for its success, in particular air assets." However, these capabilities were slow to come, and certain contingents (notably European ones[88]) already in MINUSMA continued to impose operational *caveats*, demonstrating the limits of the influence of the Security Council and France on troop-contributing countries.

Assessing the Second Phase: Cracks in the Security Council and an Uncertain Future

This second phase in the life of MINUSMA (2016–2019) marked a real turning point in France's ability to unite the Security Council and the wider "international community" behind a common strategy. The latter was initially united around the need for a more coherent mandate and in support of a Mission leadership more determined to encourage the parties to respect the 2015 peace agreement. However, the "international community" quickly became frustrated with the lack of will on the part of the Malian parties, and in particular the Malian government.

Yet France as the penholder in the Council supported Mali's request that MINUSMA adopt "a more proactive and robust approach", which shifted the problem from the Malian government's lack of political will - which France, who had its own military operation on the ground, seemed to want to spare—to pointing fingers at peacekeepers who are easy scapegoats.

Despite the unanimous vote in support of Resolution 2295 (2016), some dissenting voices, including those of Russia and some troop-contributing countries, questioned the validity of this adjustment to the mandate in a terrorist environment and highlighted the risks for

[87] Report of the UN Secretary-General on the situation in Mali, March 26, 2021, S/2021/299.

[88] European contingents from the UK, Sweden and Germany make up the Gao-based *Mobile Task Force* (MTF), which is supposed to be able to operate in all sectors as required. However, they often refuse to operate beyond a certain radius around the city of Gao and temporarily in other areas to support operations, notably because they have no confidence in the capacity of the UN's military medical evacuation systems, despite the progress made by the Mission and the UN more generally in recent years.

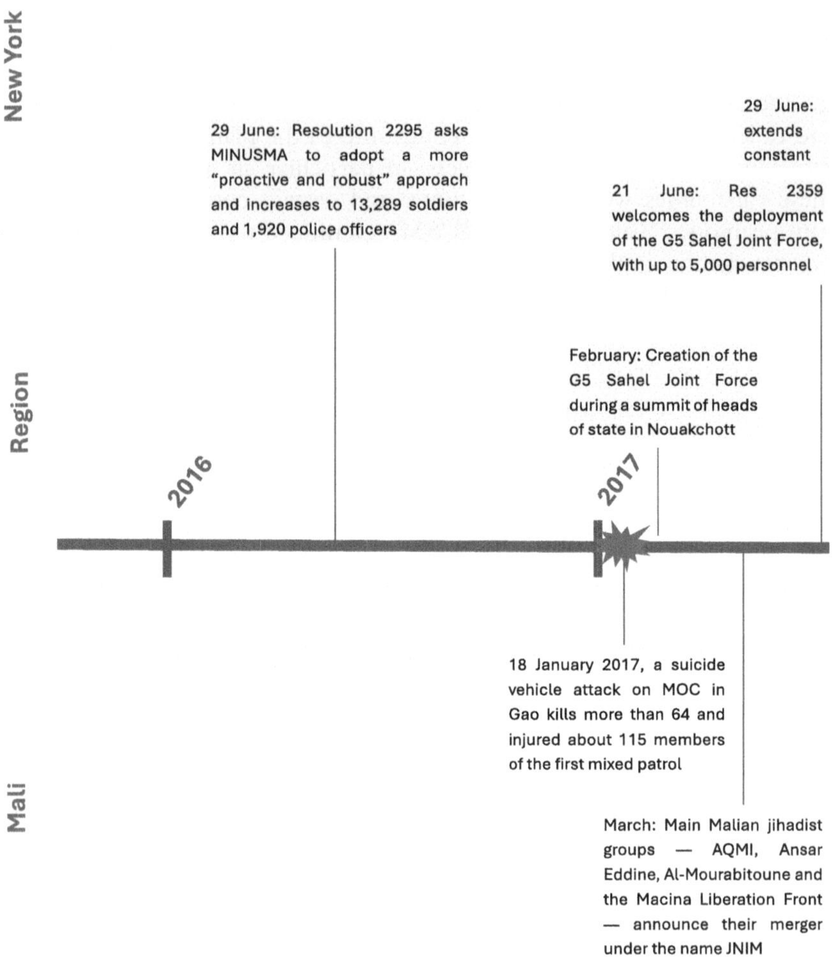

Fig. 3.2 MINUSMA timeline (2016–2019)

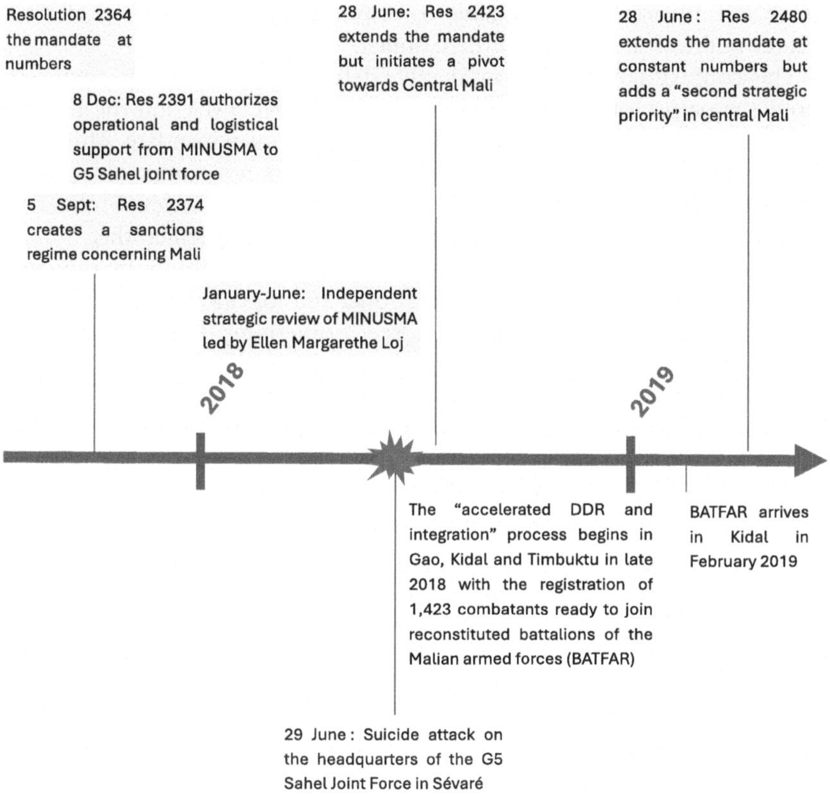

Fig. 3.2 (continued)

peacekeepers. The new responsibilities added to the mandate are the result of a mismatch between decision-makers in New York and the mission's leadership in the field. The latter was reluctant to raise new expectations to which the mission seemed unable to respond and would have preferred greater political support from the Council.

This context gave rise to considerable debates and many misperceptions about the nature of the relationship between MINUSMA and the French "parallel force" *Barkhane*, which had a counterterrorism mandate. Although the mandates and resources of the two forces were not comparable, there was inevitably a mix of genres, especially as France tried at times to influence the conduct of UN operations, even though it officially denied this.

Resolution 2391 (2017) further reinforced this ambiguity by requesting MINUSMA to also provide operational and logistical support for the joint G5 Sahel counterterrorism force, something France had long requested . It also exacerbated tensions within the Council, notably between France on the one hand and the United States and the United Kingdom on the other, over UN funding for this new regional force. The late addition of a sanctions regime at the request of the Malian government in 2017 was also divisive. Russia in particular expressed concerned about the objectivity of this regime, and France was the only country to propose names of individuals to be sanctioned.

A breaking point was reached during the controversial independent strategic review of early 2018, whose conclusions divided the UN secretariat in New York, the Mission on the ground, and the Security Council. This was for good reason. The strategic review questioned the very relevance of a peace operation in Mali given the complex and difficult realities the Mission was facing. France with the help of DPKO's leadership, once again managed to give MINUSMA a reprieve. But France also had to compromise and accept a pivot toward Central Mali and greater attention to the protection of civilians as requested by the United States, but without any additional resources, thereby forcing an *Adaptation Plan* and shifting resources away from northern Mali. The United States also introduced in the resolution the seeds of reflection on a possible future transition and eventual withdrawal of the Mission, even if France somewhat neutralized these seeds by adding criteria that made such a transition unlikely in the short or medium term (Fig. 3.2).

CHAPTER 4

MINUSMA Act Three (2020–2023): The Return of the East-West Divide and the Crisis of Host Country Consent

"Western colleagues continue to feed the Security Council with false stories about Russia's allegedly destructive activities in Mali [...] That raises the question of what their true motives are with regard to MINUSMA's activities. Despite the external pressures placed on them, we are confident that the United Nations and the Mission's leadership will remain impartial, act only in the interests of the Malian people and resolve all issues that arise through constructive engagement with the authorities of the host country."[1]

Russian representative, Anna Evstigneeva, 29 June 2022

An Exit Strategy and a Mission Put on Hold Following Two Coups d'État

The June 2020 MINUSMA mandate renewal was less controversial than the previous year's. The updating of priority measures and *benchmarks* in Resolution 2480 (2019) was not the subject of much debate. Two new priority measures were added for Central Mali in Resolution 2531 (2020) which "urges the Malian authorities to fulfil the following priority measures before the end of MINUSMA's current mandate: re-establish State presence and State authority to Central Mali ... fight against impunity for violations and abuses of international human rights law and violations of

[1] Security Council S/PV.9082, June 29, 2022.

A. Boutellis, *The UN Stabilization Mission in Mali (MINUSMA)*, Twenty-first Century Perspectives on War, Peace, and Human Conflict, https://doi.org/10.1007/978-3-031-66466-3_4

elections.[7] Discussions on a MINUSMA exit strategy were therefore put on hold as the situation in Bamako changed dramatically. The Mission was somewhat bewildered by the turn of events, to the extent that it was perceived, rightly or wrongly, as having supported President IBK since his first election in August 2013 and after his reelection for a second five-year term in August 2018.

The "international community" (represented in Mali by ECOWAS, UN, African Union, European Union, France and United States) wasted no time in unanimously condemning the military takeover and calling for the restoration of the constitutional order and the release of President IBK (regional organizations even called for his immediate return to the head of the state). France, for its part, indicated the continuation of Operation *Barkhane*. MINUSMA also continued its operations[8] and its human rights division even visited President IBK during his detention. MINUSMA head Mahamat Saleh Annadif also met him immediately after his release on 27 August[9] and engaged discrete good offices with members of the military junta.

ECOWAS, the African Union and MINUSMA, as part of the *Transition Support Group*, oversaw the establishment of the civilian transition's institutions. This led, on October 5, 2020, to the formation of a transitional government and the publication of a roadmap providing for a return to constitutional order and the holding of elections within eighteen months. ECOWAS immediately lifted the sanctions it had imposed on Mali following the coup. The Security Council, for its part, met on October 15, 2020 to adopt a presidential statement (S/PRST/2020/10) welcoming these developments and requesting MINUSMA:

> within its mandate and existing resources, to support the political transition in Mali, particularly by exercising good offices, confidence-building and facilitation at the national and local levels, and by supporting, together with the United Nations country team, the holding of inclusive, free, fair, transparent, and credible elections, conducted in a peaceful environment, includ-

[7] MINUSMA was deployed in Mali in 2013 in the context of a political transition following an armed rebellion in northern Mali and a military coup on March 22, 2012 that deposed President Amadou Toumani Touré (ATT).

[8] Le Monde, "Despite the coup d'état and Malian defiance, France maintains operation "Barkhane"," August 27, 2020.

[9] Report of the Secretary-General on the situation in MaliS/2020/952, September 29, 2020.

ing through the provision of technical assistance and security arrangements, consistent with the provisions of the Agreement. It requests the Secretary-General to include in its quarterly reports updates on how MINUSMA supports the political transition.[10]

But nine months into the transition period at the time of MINUSMA's June 2021 mandate renewal, relatively little has been achieved apart from a government action plan (PAG) and a transition government roadmap adopted by the National Transitional Council (CNT), the transition's legislative body. On April 14, 2021, the government finally announced an electoral timetable, which although respectful of the eighteen-month transition period, seemed unrealistic.[11]

Military personnel held key positions in the transitional government, including three colonels who carried out the putsch; many CNT seats were allocated to military officials, and 13 security officials were among the seventeen regional governors appointed. Although members of the transitional government are in principle not allowed to stand in the forthcoming elections, Vice-President Colonel Assimi Goïta was suspected of harboring presidential ambitions, while no candidate emerged from Mali's old political guard following the death of opposition leader Soumaïla Cissé in December 2020 from COVID-19.

On May 14, 2021, Prime Minister Moctar Ouane resigned but was immediately reappointed to carry out a government reshuffle and form a new cabinet. On May 24, 2021, the military junta however staged another coup de force, a "coup within a coup." Interim President Bah N'Daw and Prime Minister Moctar Ouane were arrested by the army and forced to resign just after announcing a new "inclusive" government. Indeed, the new government included only two key members of the military junta, who previously held the key positions of Minister of Defense and Minister of Security.[12] Despite international condemnation and the decisions of ECOWAS and the African Union to suspend Mali, the junta's strongman,

[10] Security Council, Statement by the President of the Security Council, S/PRST/2020/10, 15 October 2020.

[11] The plan is to hold a constitutional referendum on October 31 (and promulgate a new constitution on January 14, 2022), followed by presidential and legislative elections on February 27, 2022. Beyond the legal aspects of the electoral college, there are practical questions concerning the financing and securing of the electoral process in localities not controlled by the government, which remain unanswered.

[12] Crisis Group, "Maliun coup dans le coup", May 27, 2021.

Colonel Assimi Goïta, was sworn in as the new president of the transition on June 7, 2021. He appointed Choguel Maïga as Prime Minister,[13] just a few days before the renewal of MINUSMA's mandate.

France Experiences the Limits of Its Influence over MINUSMA

France, which had so far been successful in steering MINUSMA in the direction it wanted by skillfully using its position as penholder on Mali in the Security Council, was finally going to experience the limits of its influence on the ground in Mali. Paris first faced a damning UN investigation on what appeared to be a blunder by Operation *Barkhane,* and later found its historical informal control over the appointment of the head of MINUSMA challenged for the first time.

The publication by the United Nations of a report from MINUSMA's human rights division exposed the limits of France's influence on the UN mission in Mali. The report was the result of a special investigation into allegations that several civilians died as a result of an air strike carried out by Operation *Barkhane,* near the village of Bounty in central Mali on January 3, 2021. The UN report concludes:

> At the end of the investigation, MINUSMA is able to confirm the holding of a wedding celebration which brought together around a hundred civilians at the site of the strike, including five armed persons, presumed members of the Katiba Serma. At least 22 people, including three of the presumed members of Katiba Serma present at the gathering, were killed by the *Barkhane* Force strike on 3 January 2021 in Bounty [...] The group affected by the strike was overwhelmingly made up of civilians, who are protected from attack under international humanitarian law.[14]

On the French side, it was the Ministry of the Armed Forces, not the Ministry of Foreign Affairs, which reacted strongly to the content of the MINUSMA report in a lengthy press release on 30 March. While taking care to "welcome the fact that a UN mission can investigate in complete independence [...] an essential condition for the vitality of values and respect for democratic principles to which France is firmly attached", the

[13] Marking the M5-RFP's entry into the transitional government.

[14] MINUSMA, Human Rights and Protection Division, Report on the Bounty incident of January 3, 2021, March 2021.

Ministry vehemently opposed the conclusions of the UN investigation. In particular, it questioned the methodology of the UN investigation, and reiterated the French line of defense, unchanged since early January, that the strike did indeed target an armed terrorist group "identified as such" following "a robust process" of targeting and "strict and standardized procedures", themselves enshrined in "respect for the law of armed conflict."[15]

The French Ministry could have left it at that. However, it went on to place the UN report in the context of an "information war" which France increasingly felt the target of in the Sahel, due in particular to the growing appetites of Russia, Turkey and China in the region, but also to a rise in anti-French sentiment. In fact, this issue was at the heart of the January 2020 Pau Summit, which brought together France and the G5 Sahel countries. This same idea of "an informational battle" was later echoed in a public report by the French parliament on Operation *Barkhane*, which argued that "France is not bogged down in the Sahel", and that its record there is "unquestionably positive."[16]

On 1st April, on a trip to Bamako, the French Minister of the Armed Forces, Florence Parly, even casted doubt on the impartiality of MINUSMA (an "independent investigation, at least I hope so"), reiterating the argument of "information warfare" ("our enemy exploits every possible polemic"). Majority MP Fabien Gouttefarde (LREM) even wrote on *Twitter*: "I was at this meeting. And I reiterate my support for our troops in the Sahel, who respect the laws of war. I regret that the UN is in turn the vector of an informational war being waged against France and against *Barkhane*."[17] In an interview with the French daily *Le Figaro,* General Lecointre, Chief of Staff of the French Armed Forces stated that "very clearly, this was an attack on the French army, on Operation *Barkhane,* against the legitimacy of our commitments", and that the "errors" and "biases" in the MINUSMA report "stem in part from manipulation", before concluding: "I think that we will henceforth be systematically confronted with this type of attempt to hinder us, to discredit us, to

[15] Elise Vincent and Cyril Bensimon, "L'armée française aurait bien commis une bavure au Mali selon un rapport de la MINUSMA", *Le Monde*, March 31, 2021.

[16] Assemblée Nationale, Rapport d'information déposé par la commission de la défense nationale et des forces armées en conclusion des travaux d'une mission d'information sur l'Opération Barkhane N° 4089, April 14, 2021.

[17] Rémi Carayol, Bounti's blunder boils down to an "informational battle" for France, April 14, 2021.

delegitimize our action, to turn the population against our action."[18] The researcher Yvan Guichaoua notes:

> In the past, this division [MINUSMA's human rights division] may not have made civilian casualty files public (only to use them as negotiating leverage with armed groups), but in the case of Bounty I can see a clear piece of work. The investigation was announced, and the report made public within a very reasonable timeframe. This is the first time that MINUSMA has put so many resources into an investigation, with a detailed, serious report, precisely because it was obvious that the slightest flaw would be exploited. Despite this, France is putting up a front of denial [...] There is a raison d'État here, the image of France must be preserved at all costs.[19]

The tensions between France and MINUSMA over the Bounty report were symptomatic of France's diminishing influence in Mali at a time when France was already perceived as preparing to downsize *Barkhane* (at the same time the US army had begun the final phase of its withdrawal from Afghanistan). The publication of the MINUSMA Bounty report interestingly took place at a time of transition and vacancy at the head of the Mission. Special Representative of the Secretary-General Mr. Annadif had left the country and his successor El-Ghassim Wane had not yet arrived. But it was also an opportunity for MINUSMA, which had long been perceived as "France's tool", to assert a more independent and impartial voice. France's harsh declarations had the effect of upsetting even its best allies and relays[20] within the Mission and thus led to a loss of French influence within MINUSMA.

At the June 2021 MINUSMA annual mandate renewal, France was now facing a new American administration under President Joe Biden which it thought could be more supportive. It explained that, "MINUSMA's resources remain limited in terms of addressing the situation in central Mali. That priority was conferred upon MINUSMA in 2019 without providing it with the needed additional resources. The Mission has only 2,000 peacekeepers and three helicopters at its disposal every two weeks in the area. We therefore would like to seriously propose an increase in resources

[18] Jeune Afrique, "Mali UN report on Bounti is 'an attack on Barkhane' according to France," May 22, 2021.

[19] Yvan Guichaoua: "There's a raison d'état in Mali a raison d'État, the image of France must be preserved at all costs" *Middle East Eye*, April 6, 2021.

[20] French employees, including former military personnel, but not only.

and the troop ceiling, which would enable the Mission to protect civilians more effectively and promote reconciliation among communities without jeopardizing its operations in the rest of the country. Such reinforcement would complement the implementation of the adaptation plan."[21]

This time, the United States was not opposed to the idea, but asked for more time to submit the proposal through U.S. Congress due to its budgetary approval process. But African members of the Council (Kenya, Niger, Tunisia and Saint-Vincent-et-les-Grenadines) expressed doubts about an increase in troop numbers, concerned about the possible relation to President Macron's announcement a few days earlier that it was withdrawing Operation *Barkhane*, and that France may be seeking to let the UN handle the consequences of its withdrawal. For its part, China questioned the validity of the French request, implicitly suggesting that France was seeking to alter the UN recommendations:

> when the Secretary-General submitted his report (S/2021/519) to the Council on 1 June, he already recommended that the ceiling of the Mission's uniformed personnel be maintained. In China's view, the set recommendation contained in Secretary- General's report is in line with the current situation in Mali and the needs of the Mission in performing its mandate. We would like to express our full support for that. China questions the need for the resolution to request the Secretary-General to submit another report on the same issue after merely a month or so.[22]

France finally withdrew its proposal to increase the troop ceiling, and instead included in the resolution a request to the Secretary-General "to provide no later than July 15, 2021, a report on the progress of the Force Adaptation Plan implementation and recommendations on the force levels and ceiling of MINUSMA uniformed personnel, including planning assumptions and estimated costs."[23]

Resolution 2584 (2021) was finally adopted unanimously, extending the mandate but at constant troop levels, while adding to MINUSMA's main strategic priority (which remained supporting the implementation of the Agreement by the Malian parties) the support for the political

[21] Security Council Seventy-sixth year, 8794th meeting, Monday, June 14, 2021, 10 a.m., New York. S/PV.8794. p7.

[22] Security Council, S/PV.8809, June 29, 2021. p5.

[23] Security Council Resolution 2584, S/RES/2584 (2021), June 29, 2021.

transition already agreed in the October 2020 Council President's statement.[24] The United States representative said he was "grateful for France's collegiality in developing resolution 2584 (2021) and in conducting negotiations," but new, more "classic" fault lines between East and West were already emerging, even if they were for the moment expressed primarily around the apparently peripheral issue of climate-related security risks.[25]

The limits of French influence also began to be felt in MINUSMA's senior leadership. The Mission had had five heads and special representatives of the Secretary-General between 2013 and 2022, almost all of whose names were put forward by Paris for appointment by the Secretary-General. Bert Koenders from the Netherlands helped position the Mission as a key player in the first phases of the Algiers negotiations in 2014, alongside Algeria, the lead country in international mediation, and the African Union.[26] MINUSMA's Deputy Special Representative in charge of Political Affairs, Arnauld Akodjenou, a binational Franco-Beninese, then served briefly as interim head. In December 2014, the Tunisian Mongi Hamdi was appointed, but only stayed for a year before being encouraged to resign in December 2015.[27] Algeria, which had initially suggested Mongi Hamdi's name, no longer supported him, which enabled France to suggest a replacement in the person of Chad's Mahamat Saleh Annadif, previously his country's Minister of Foreign Affairs (from 1997 to 2003).

Annadif remained at the head of the Mission for five years, until March 2021, and made a lasting mark on the Malian peace process, through the use of his good offices to put an end to clashes between armed groups, as well as his more discreet good offices to try to prevent and resolve political crises in Bamako. Unlike his predecessor, Mahamat Saleh Annadif already benefits from extensive experience of peace processes and the stature of his country Chad, a regional military power contributing 1400 blue helmets

[24] Security Council, Statement by the President of the Security Council, S/PRST/2020/10, 15 October 2020.

[25] Security Council, The situation in Mali S/PV.8809, June 29, 2021.

[26] He was previously head of the UN mission in Côte d'Ivoire and will leave MINUSMA to become Foreign Minister of the Netherlands.

[27] During this period, MINUSMA lost some of its influence and it was at this key moment that Algeria decided to take the lead in implementation, and that it would chair not only the Comité de Suivi (CSA) but also all its sub-committees in the agreement's implementation phase, contrary to established practice when a Security Council-mandated mission is present on the ground.

to MINUSMA and one of the countries in the region invited by Algeria to sit on the CSA. With his knowledge and understanding of armed groups, Annadif was able to establish good relations with all signatories to the agreement, both government and armed groups.

On the eve of his departure, and against a backdrop of French concerns about growing Russian activism in Africa, Annadif symbolically gave an exclusive interview to RT France (although he gave his last interview in early May 2021 to *France24*). In this interview, he insisted that the Security Council was united behind MINUSMA and that "questioning the mandate is not on the agenda." In Mali, he insisted on the need for a "national awakening of the political class." While acknowledging "*Barkhane*'s success" in neutralizing jihadist leaders, he concluded that, "all wars, all battles have always ended in discussions, in negotiations [...] beyond nuances, as soon as there is a will to make peace, the hand must be extended [...] peace must be made with those who are willing to make it."[28]

The succession of Mahamat Saleh Annadif as head of the Mission in Mali—discussed since the start of 2020—brought new tensions between Paris and Washington. Donald Trump's administration apparently supported the candidacy of the American former deputy head of mission David Gressly.[29] France, for its part, insisted that another French-speaking African should succeed Annadif, and won that battle again. Following the election of Joe Biden the US administration conceded to France the appointment of Mauritanian El-Ghassim Wane[30] to head MINUSMA.

Even though his room for maneuver was reduced by divisions in the Security Council and above all the tensions between France, the penholder on the one hand, and the Malian transitional government and its new Russian ally on the other, the facts that Mr Wane is African and had previously served at the African Union were undoubtedly assets. In the past, the Malian host state had not hesitated to target a French employee of MINUSMA to stir up anti-French sentiment in Mali. In December 2019, the Malian government

[28] RT France "The Big Interview: Mahamat Saleh Annadif", November 27, 2019.

[29] He was previously DSRSG/RC/HC of MINUSMA and more recently of MONUSCO in the DRC.

[30] El-Ghassim Wane returns to the United Nations where he served as Assistant Secretary-General in the Department of Peacekeeping Operations (2016–2017) after a career with the African Union (AU), where he was Chief of Staff and Senior Advisor to the Chairperson of the AU Commission, Moussa Faki Mahamat (from 2017 to 2019) and previously the Commission's Director of Peace and Security and Head of the Conflict Management Division.

expelled the head of the MINUSMA office in Kidal, Frenchman Christophe Sevillon, on the pretext that he had welcomed delegations "from Mali and abroad" to Kidal during the MNLA congress, which was misinterpreted by the government delegation on site.[31] Around the same time MINUSMA spokesman Olivier Salgado, another Frenchman, came very close to suffering the same fate, following violent comments by Malians on social networks.[32] Olivier Salgado was eventually expelled later in July 2022 by the ruling junta, which accused him of having posted "unacceptable information" on Twitter about the case of the 49 Ivorian soldiers detained in Bamako.[33]

Despite some United States' attempts to impose its views here and there, France ultimately kept the lead in the Malian dossier. France's penholder role and a certain division of labor within the Council according to the particular interests and zones of influence of the major powers were maintained. But as relations between Paris and Bamako deteriorated rapidly during the second half of 2021 and throughout in 2022, against the backdrop of the deployment of Russian mercenaries from the *Wagner Group* and verbal escalation, MINUSMA and its chief El-Ghassim Wane nevertheless found themselves in a very delicate situation.

Franco-Malian Divorce and Power Rivalries Against a Backdrop of *Wagner*

At the end of summer 2021, the Reuters news agency revealed that the Malian transitional authorities were close to concluding an agreement with *Wagner Group* to send around a thousand Russian mercenaries to Mali.[34] Earlier, President Emmanuel Macron had announced the reconfiguration of Operation *Barkhane* in the Sahel.[35] France tried to dissuade

[31] Aïssatou Diallo, "Mali: expulsion of Frenchman Christophe Sivillon, head of the Minusma office in Kidal", Jeune Afrique, December 10, 2019.

[32] Olivier Salgado had stated that the mission had made arrangements "so that the MINUSMA head of office could leave Kidal and Malian territory as soon as possible, in the interests of the parties." Bamada.net, "Après l'expulsion du représentant de la MINUSMA à Kidal: à qui le tour?" December 12, 2019.

[33] France 24, "Mali expels Olivier Salgado, Minusma spokesman", July 7, 2022.

[34] John Irish, David Lewis. "Mali close to an agreement with Russian mercenaries of the Wagner group", September 13, 2021.

[35] with France's military footprint reduced by half by early 2022 to around 2500 to 3000 soldiers, and the closure of certain French bases in northern Mali French forces will continue to target the leaders of groups affiliated to the Islamic State (EI) and al-Qaeda, and to support local armies, notably through Task Force Takuba (made up of European special forces), whose headquarters will be moved from Mali to Niger.

the Malian junta from calling in the Russian mercenaries, multiplying its declarations and official visits to Bamako. Paris warned its Malian partners that such an arrangement would be "incompatible" with maintaining the French military presence in Mali. Germany with almost 1500 soldiers deployed in Mali as part of MINUSMA, also indicated that any agreement between the ruling junta in Bamako and the private Russian military company *Wagner Group* would "call into question" the German army's mandate in Mali.[36]

At the United Nations General Assembly in New York on September 25, 2021, Mali's transitional Prime Minister, Choguel Maïga, harshly criticized France suggesting that "the new situation created by the end of *Barkhane* placing Mali before a fait accompli and exposing it to a kind of abandonment in mid-air, leads us to explore ways and means of better ensuring security autonomously with other partners." He deplored a "lack of consultation" on the part of Paris and a "unilateral" announcement without tripartite coordination with the UN and the Malian government. And like previous Malian governments before, he also called for "a more offensive posture" on the part of MINUSMA peacekeepers.[37]

Despite Malian steady denials, Russian Foreign Minister Sergei Lavrov confirmed in a press conference the Malian authorities' request to the "private Russian military company", while denying any Russian government involvement. A few days later, Mali's transition Minister of Defense, Colonel Sadio Camara, took delivery of four Mi-171 helicopters (as part of a contract signed in December 2020 and that took effect in June 2021), as well as weapons and ammunition donated by Russia, in front of the cameras.[38] Two more Russian MI-35P attack helicopters and a radar system arrived in Mali in April 2022,[39] followed by five L39 and Sukhoi 25 fighter jets in August 2022.[40]

[36] AFP, "Wagner In Mali Après La France L'allemagne S'offusque", VOA, September 15, 2021.

[37] Le Monde, "Mali reproche à la France un "abandon en plein vol" dans la lutte contre les Djihadistes au Sahel", September 26, 2021.

[38] TV5Monde, "Military cooperation: Mali receives four Russian helicopters and praises partnership with Moscow." October 1, 2021.

[39] Twitter, "The Chief of the General Staff of the Armed Forces, Major General Oumar Diarra received this Monday on behalf of the Minister of Defense and Veterans, 2 combat helicopters and surveillance radars," Presidence Mali @PresidenceMali, April 18, 2022. https://twitter.com/PresidenceMali/status/1515976156662710272.

[40] TV5Monde, "Mali: la junte reçoit de nouveaux équipements militaires de la Russie", August 9, 2022.

Clearly upset, France was keen to highlight the inconsistencies in the position of Mali's transitional Prime Minister, Choguel Maïga. It publicly circulated a letter from its ambassador in Bamako addressed to Choguel Maïga and dated September 24, 2021, which revealed that "Mali's current opposition to the increase of the contingent [MINUSMA force by around 2000 Blue Helmets, recommended earlier by the UN Secretary-General][41] has resulted in the immediate withdrawal of the planned vote"[42] on a new Security Council resolution initiated by France after all the diplomatic efforts, notably with the United States detailed in the previous section. Ultimately, it was the Malian transitional authorities who blocked the French proposal "to deploy 2,069 additional uniformed personnel, including 1,730 military personnel, 300 members of formed police units and 39 police officers outside formed units,"[43] as they seemed to (continue to) believe that these additional Blue Helmets would not be useful unless they have a truly offensive mandate:

> It was in the Security Council chamber that our ambassador discovered a draft resolution calling for an increase of more than 2,000 men in the strength of MINUSMA. Our ambassador said that his government was not aware of this. As we say in Africa, you can't shave the heads of Malians in their absence. We can't decide to send a military contingent to our territory without knowing its origins, missions, strength, nationalities and objectives. So we opposed it.[44]

The intention of MINUSMA's leadership to use this increase in the troop ceiling as leverage to encourage greater political involvement on the part of the Malian authorities was therefore neutralized by the growing tensions between the French penholder and the Malian transitional

[41] Le Figaro, "Mali UN chief recommends increasing Minusma by around 2000 blue helmets." July 16, 2021.

[42] Letter from the French Ambassador addressed to Mali's transitional Prime Minister, Choguel Maïga, September 24, 2021, circulating on social networks.

[43] Security Council, Report of the Secretary-General, "Adaptation and authorized strength of the United Nations Multidimensional Integrated Stabilization Mission in in Mali" S/2021/657, July 16, 2021.

[44] "Malian Transitional Prime Minister, Choguel Kokalla Maïga: "La France" exerce un Terrorisme Politique, Médiatique et Diplomatique Contre notre Gouvernement." Anadolu Ajansı.

authorities.[45] France was more openly assuming its influence strategy, that is its ability to influence the behavior of other states in line with its interests,[46] including the French army's "informational struggle"[47] but a little late in the game.

The verbal escalation between Paris and Bamako continued when President Macron denounced Choguel Maïga's undiplomatic remarks at the UN. He spoke of "a disgrace, which dishonors what is not even a government"[48] and then declared that, "the French army has no business substituting itself for the non-work of the Malian state." As a result, Malian Foreign Minister Abdoulaye Diop summoned the French ambassador in Bamako, and denounced the "unfriendly" and "derogatory" remarks, "likely to damage" relations between the two countries.[49] Mali's transitional Prime Minister Choguel Maïga reiterated his criticism a few days later in an interview in the French newspaper *Le Monde* dated October 18, 2021:

> The facts are that Mali asked France to help destroy terrorism and recover its entire territory. Almost nine years later, what have we seen? Terrorism, which was confined to Kidal, has spread to 80% of our territory. This leads Malians to believe that there is an international plot against our country […] Last June, we woke up one morning to hear in the media that France was suspending its military operations with the Malian army, without warning us or giving us any explanation, all because a new government that did not suit them had been put in place (following the second coup d'état on May 24). A month later, at the G5 Sahel summit, Emmanuel Macron came to announce that *Barkhane* was leaving […] "Since then, our government has clearly understood that if it relies on just one partner, it could be abandoned at any time. We're looking for others […] I don't know any *Wagner*. These are just rumors at this stage. The day we conclude agreements with any

[45] In December 2021, the Malian Ministry of Foreign Affairs announced that it had agreed to the deployment of 1000 additional Chadian soldiers "within a bilateral framework and at the request of the government of the Republic of Chad […] to face threats and protect its troops […] in northern Mali". […] to meet the threats and protect its troops […] in northern Mali following the reconfiguration of the Barkhane force" although the Security Council has not yet has yet to approve the necessary increase in MINUSMA's troop ceiling. "Le Tchad Envoie 1000 Soldats Pour Renforcer La Mission De L'onu Au Mali", RFI, 2021.

[46] Ministry of Europe and Foreign Affairs, "Roadmap of influence" December 2021.

[47] Manon Laplace, "Mali quand la France lance dans la guerre de communication", Jeune Afrique, November 16, 2021.

[48] RFI, "Mali first political reactions to French president's remarks." October 2, 2021.

[49] RFI, "Mali the French ambassador summoned by the authorities", October 6, 2021.

country, we'll make them public. In the meantime, don't accuse us of intentions.[50]

France which had seemed somewhat isolated in its verbal escalation, started to use its influence more effectively with its Western and African allies, mobilizing them in a sacred alliance against the arrival of Russian mercenaries in Mali. The US State Department issued a communiqué[51] on 15 December 2021, followed by a Joint Communiqué by Europeans (plus Canada) denouncing the *Wagner Group*.[52] The US communiqué directly referred to the Malian authorities' refusal to accept the increase of 2000 blue helmets proposed a few months earlier by the UN Secretary-General:

> Furthermore, the invitation to *Wagner Group* would disrupt efforts by the international community to support the fight against terrorism and could put at risk the contributions of more than 20,000 international peacekeepers and troops who serve Mali at no cost to the people or government. The United States laments that Mali's transitional government has refused to accept more than 2,000 additional MINUSMA military and police peacekeepers — again, at no cost to Mali — that would have contributed to civilian protection. The United States also regrets that the transitional government has delayed efforts by international partners to deploy additional troops and trainers and to bolster security operations. We call on the transitional government to take action to facilitate responsible and accountable security assistance efforts aimed at protecting and empowering the Malian people.[53]

At the time of President Macron's announcement on June 11, 2021 that the *Barkhane* force would be withdrawn,[54] France seems to have neglected the UN operation in its communication—even though it had just tried to increase the number of peacekeepers, and despite the fact that

[50] Morgan Le Cam, "Choguel Maïga: "Notre gouvernement a compris que s'il ne compte que sur un seul partenaire, il pourra à tout moment être abandonné"," Le Monde, October 19, 2021.

[51] Suggesting that, "*Wagner*'s forces forces, known for their destabilizing activities and human rights violations, will not bring peace to Malibut will further destabilize the country."

[52] Ministry of Europe and Foreign Affairs, "Joint communiqué on the deployment of the Wagner group in Mali" December 23, 2021.

[53] U.S. Department of State Office of the Press Secretary, "Press Release: Potential Deployment of the Wagner Group in Mali" December 15, 2021.

[54] Le Parisien, "Emmanuel Macron announces the end of Operation Barkhane in the Sahel", June 10, 2021.

MINUSMA would remain the only international force present in Timbuktu, Kidal, and Tessalit. At the time of the handover, however, MINUSMA's role was more clearly recognized by the French military. On 12 October, the day the last French military convoy left the Kidal base (leaving behind 1300 Guinean and Chadian UN peacekeepers[55]), the spokesman for the French military declared:

> The priority today is not to regain control of the whole region, but to contain the threat by holding key points where the population and communication routes are located. Malian forces have been trained for this. MINUSMA, for its part, must play its peacekeeping role. France will continue to carry out targeted operations from a distance. For the rest, a foreign military presence cannot be a solution. This is a national matter that only Mali can deal with.[56]

Another December 2021 communiqué from the French Ministry of the Armed Forces on the transfer of the Timbuktu camp stated:

> As in Kidal and Tessalit, this transfer has been gradual and closely coordinated with the Malian Armed Forces, which have a strong garrison in Timbuktu, and with the United Nations Multidimensional Integrated Stabilization Mission in Mali MINUSMA), which has nearly 2,200 soldiers permanently deployed there.[57]

Against this backdrop of high tensions between the ruling Malian junta and Western countries, President Macron briefly envisaged a visit to Bamako on 20 December 2021, before retracting his plans, citing the health situation as an excuse. The French president's retraction was also motivated by Bamako's refusal to accept the Élysée's organization of a mini-summit bringing together the current presidents of ECOWAS (Ghana) and the G5 Sahel (Chad)—aimed at putting pressure on the Malian interim president, Colonel Assimi Goïta.[58] The French ambassador

[55] David Baché, "MaliMINUSMA soldiers apprehensive about Barkhane withdrawal" RFI, June 15, 2021.

[56] David Rich, "The end of Operation Barkhane What strategy for French forces in the Sahel? 24," October 14, 2021.

[57] Philippe Chaplean, "After Kidal and Tessalit, Timbuktu in turn handed over by the Barkhane force," Ouest France, December 14, 2021.

[58] Nicolas Beau, "Emmanuel Macron's incomprehensible trip to Mali December 20", Monde Afrique, December 15, 2021.

to Bamako was declared persona non grata on January 31, 2022, after French Foreign Minister Jean-Yves Le Drian spoke of an "illegitimate junta."[59]

Return of an East-West Security Council Divide

The Security Council's visit to Mali (the fifth since the start of MINUSMA) in late October 2021 at the initiative of France jointly with Niger and Kenya, had no particular effect. Not only did the Malian transitional authorities give no assurances as to their willingness to present an electoral timetable in line with the demands of the "international community," but in addition, the day after the Council visit, they expelled the ECOWAS representative from Mali.[60]

Both ECOWAS and France found themselves in an awkward position, given their inconsistent political approaches to military coups in Sahelian states. As Niagalé Bagayoko (president of the *African Security Sector Network*) pointed out: "France has never managed to reconcile the defense of its interests on the one hand, and its values on the other. The result is an ambivalence that makes its position in the Sahel untenable. In the name of stability, it supported the coup d'état in Chad but is opposed the coups in Mali and Guinea."[61] She explains:

> I think that what France defends in Mali is its status as a power. It is seeking to demonstrate that it still has the foreign policy instruments, starting with the military instrument, to influence international relations. And unfortunately, the failure it is currently experiencing, like all the players in the Sahel, and indeed all the players involved in this type of insurrectionary environment, tends to call into question this status of power to which it claims to be entitled.[62]

[59] Carayol, Rémi. "In MaliFrance En Terre Hostile." Mediapart. Mediapart, February 2, 2022.

[60] "Security Council Returning from Mali and Niger et du Niger, une mission du Conseil de sécurité présente un bilan contrasté de la situation dans ces deux pays: communiqués de presse," CS/14683, October 29, 2021.

[61] Rémi Carayol, "Barkhanele crépuscule de la force", Mediapart, November 29, 2021.

[62] TV5Monde, "MaliNiagalé Bagayoko: Un mouvement populaire massif pourrait forcer la France to leave the country", January 16, 2022.

At the same time, the states of the region, and Mali in particular in particular, well aware of the power rivalries between members of the Security Council took advantage of it. What at first sight appeared to be a success for Paris in asserting its power and its influence on its allies in the region, nevertheless put ECOWAS leaders in an awkward position vis-à-vis their increasingly anti-French public opinions. The latter were also very of ECOWAS and its sanctions against the military junta who came to power through coups, which also affected the general economy of these countries.[63] Two years later on January 28, 2024, the three military juntas in power in Mali, Burkina Faso and Niger would ultimately announce that there three countries are leaving ECOWAS altogether.

On February 17, 2022, France and its European partners formalized their military withdrawal from Mali[64] justifying it by "multiple obstructions from the Malian transitional authorities." French President Emmanuel Macron at a press conference alongside the Senegalese and Ghanaian presidents, added, "We cannot remain militarily involved alongside de facto authorities whose strategy and hidden objectives we do not share," and who have recourse to "mercenaries from the [Russian] company *Wagner.*".[65]

The French withdrawal would take four to six months and "during this period, we will maintain our support missions for MINUSMA," adds President Macron, suggesting a closeness between France and the UN mission at a time when MINUSMA was precisely seeking to reassert its impartiality. These remarks also suggested that MINUSMA depended on support from *Barkhane*. This led to criticism on social networks that the UN Mission was unable to protect itself.[66] With the French departure, social media networks seem to have refocused their attention on MINUSMA, which had previously only been a secondary target of online attacks including misinformation and disinformation.

[63] Sometimes lax in the face of problematic (re)elections (including "third terms" in Côte d'Ivoire and Guinea), and at other times firm in the face of coups d'état in Bamako and Conakry which, although led by the military, removed elected presidents from power when they became highly unpopular. Le Monde, "Au Sahel, un sentiment anti-français exacerbé", Monde Afrique, December 7, 2021.

[64] Of the remaining Barkhane bases in Gao, Gossi and Ménaka, including Task Force Takuba.

[65] France24, "France and its partners announce their military withdrawal from Mali," February 17, 2022.

[66] "France et Ses Partenaires Annoncent leur Retrait Militaire du Mali". 2022. *France 24.*

France as the penholder in the Security Council on Mali, theoretically the leading defender of MINUSMA, had become a little cumbersome as a result of its deteriorating relationship with the host state and Russia. As the French newspaper *Le Monde* remarked, "Russia's weight on the UN Security Council is reducing the room for maneuver of the UN Special Representative for Mali, El-Ghassim Wane. At the presentation of his quarterly report on January 11, 2022 in New York, he gave an alarmist assessment of the security situation in Mali, without ever uttering the word *Wagner*."[67] As in the case of the Central African Republic, the UN Secretariat's reports on Mali never mention the Russian mercenaries *Wagner Group* by name.

At the next Security Council session on Mali on 7 April 2022, the return of classic East-West divisions was even more obvious. On February 24, 2022, Russia launched an illegal military offensive in Ukraine and Council debates henceforth became tainted by the invasion. Around the same time as the Boutcha massacre in Ukraine, a massacre took place in Moura in Central Mali, 50 km away from the town of Mopti. There several hundred civilians were allegedly killed at the end of March 2022 during a joint military operation by Russian *Wagner Group* mercenaries and Malian armed forces.[68] In the Council, Russia supported by China opposed a French text calling for "thorough and independent investigations to establish the facts, find those responsible for these violations and exactions and bring them to justice." The United States directly implicated Russia, and warned Mali and African countries in general "against any partnership with the *Wagner Group*, which is linked to the Kremlin":

> *Wagner* forces are implicated in human rights violations, including summary executions in the Central African Republic and elsewhere. The *Wagner* forces will not bring peace to Mali divert resources from the Malian Armed Forces' fight against terrorism and ultimately undermine the stability of the region.[69]

[67] Bobin, Frédéric, Cyril Bensimon, Jean-Philippe Rémy, Elise Vincent, Benoît Vitkine, and Eliott Brachet. "Du Sudan au Mali comment la Russie place ses Pions en Afrique." Le Monde, January 28, 2022.

[68] France 24, "Massacres at Moura in central Mali: les dessous d'une opération militaire," April 6, 2022.

[69] Security Council, S/PV.9012, April 7, 2022.

The Russian Federation unreservedly supported the Malian transitional government and pointed the finger of blame at the West, not hesitating to use its relationship with the Malian host state to settle scores with Western countries, starting with France. Russia denounced the fact that the reduction in France's military presence "creates a security vacuum" and deemed the French decision to completely withdraw from *Barkhane* "irresponsible" at a time when Westerners "are paying the price for their 2011 intervention in Libya." It stated that it "provides assistance to the Malian Armed Forces under bilateral agreements," and denounced what it described as a "disinformation campaign about so-called Russian mercenaries," and part of an "unscrupulous geopolitical game." Finally, Russia denounced sanctions against Mali as well as against Russia which it argues demonstrates the importance for them to defend their independence and overcome the "consequences of colonialism."[70]

Russia's Deputy Permanent Representative to the UN, Anna Evstigneeva, concluded her April 2022 address to the Council with a scathing:

> In the initial version of my statement, I expressed the hope that Mali's traditional partners would abandon their self-righteous superiority and the language of sanctions, and that, in a mutually respectful dialogue with Bamako, they would find ways of ironing out differences. However, in view of the present discussion, I have to admit that this hope is unlikely to materialize, and that Western states will not heed the call for equal, mutually beneficial and respectful cooperation.[71]

The rhetoric of the various members of the Security Council had changed. On the one hand, France and the United States no longer talked about Mali only. They now directly implicated Russia and warned African countries as a whole against Russia's influence offensive on the continent through the *Wagner Group*. On the other hand, Russia and its Chinese ally now blocked the texts proposed by France as the penholder, giving their unreserved support to the transitional host government.

[70] United Nations, "Security Council Special Representative for Mali calls for redoubled efforts to reach agreement on transition," CS/14856, 9012nd meeting, April 7, 2022.

[71] Security Council, S/PV.9012, April 7, 2022. P19.

MINUSMA Caught in the Crossfire Between Council Members and the Host State

The entry of Russia and *Wagner Group* mercenaries onto the Malian scene, in addition to escalating tensions between the Malian transitional government and France would quickly have repercussions for the UN mission on the ground. The consent of the Malian host state crumbled. The Malian ruling junta, emboldened by increasingly assertive Russian support, played the populist card and stirred up anti-French sentiment, but also started sending signals to the UN Mission on the ground that its services may no longer be needed. A first indication came from the fact that in December 2021 the Malian armed forces stopped requesting medical evacuation support (*medevac*) from MINUSMA.

In mid-January 2022, after Paris suspended *Air France* flights to Bamako on 12 January,[72] MINUSMA had to cease its own flights for several days, after the junta formally protested against UN aircraft overflying the *Sofara* military camp—where Malian special forces and *Wagner Group* mercenaries were based—in Central Mali.[73] MINUSMA flights eventually resumed, but only after the UN acquiesced to a Malian request to submit formal requests and flight plans 48 to 72 hours in advance, including for observation drones. Mali also imposed extensive no-fly zones for MINUSMA (and *Barkhane*) in central Mali and the region south of Ménaka. During the last quarter of 2022, the UN reported that, "237 MINUSMA flight requests have been either denied or received no response from Malian authorities," who deplore the fact that the mission "despite the commitments made, does not share the information and data collected by the drones."[74]

A succession of events made the rift between MINUSMA and the Malian host state widen even further. On March 22, 2022, a serious and unprecedented incident took place. A Malian attack helicopter (probably piloted by Russian paramilitaries) fired six rockets in the direction of British peacekeepers in the *Tessit* region, one of which landed just 50 meters from MINUSMA vehicles. MINUSMA did not immediately report

[72] France 24, "Mali denounces a 'violation' of its airspace by a French military aircraft," January 13, 2022.

[73] Bensimon, Cyril, and Elise Vincent. "In Mali Les Mercenaires Wagner Progressent Au Centre Du Pays." Le Monde.fr. Le Monde, January 6, 2022.

[74] Report of the Secretary-General, S/2023/36, January 16, 2023.

the incident publicly, but at the subsequent Security Council meeting on April 7, 2022, the United States revealed the incident:

> the United States is deeply concerned by the Malian authorities' disregard for the status-of- forces agreement with MINUSMA and the restrictions on the Mission's movements, especially in central Mali. The 22 March incident involving the Malian forces firing six rockets near peacekeepers in eastern Mali is unacceptable. It is an affront to all those who serve in United Nations peacekeeping missions. The transitional Government is responsible for ensuring the safety, security and freedom of movement of United Nations personnel and assets. Restrictions limit MINUSMA's ability to anticipate, deter and effectively respond to threats to civilians, including from violent extremist groups, and will only lead to more attacks against civilians and Malian forces. These restrictions must end so that MINUSA can fulfil its mandate to protect civilians and foster peace.[75]

In a separate incident, the host state did not allow MINUSMA human rights investigators to visit the site of the alleged massacre of several hundred civilians in Moura (central Mali) during a joint military operation by *Wagner Group* mercenaries and Malian soldiers in late March 2022. It was only on May 12, 2023 that MINUSMA finally published the results (alleging the killing of "at least five hundred individuals, including around twenty women and seven children",[76] when Malian authorities claim to have "neutralized" 203 jihadists) of a difficult Moura investigation. MINUSMA had been prevented from visiting Moura in following weeks despite pressure from a number of Western countries,[77] UN investigators were arrested and detained for several hours at the end of April 2022 while meeting Moura survivors in the neighboring town of Sévaré,[78] and MINUSMA international staff involved in the investigation ultimately had to be moved out of the country due to repeated intimidations by Malian security services.

Human rights reporting by the UN clearly became a strong irritant for the Malian transitional government and its Russian partner. On February 5, 2023, the government expelled the director of the MINUSMA human

[75] Security Council, S/PV.9012, April 7, 2022.

[76] OHCHR, Rapport sur les évènements de Moura du 27 au 31 mars 2022, May 2023.

[77] Mali authorized a reconnaissance overflight by a MINUSMA helicopter on 3 April but no ground investigation. France 24, "Massacres at Moura in central Mali: the underside of a military operation," April 6, 2022.

[78] RFI, "La Minusma constate une explosion des violences au Mali", May 30, 2022.

rights division, Guillaume Nguefa,[79] days after violently denigrating the speech of a Malian human rights advocate who had apparently been invited by an elected Council member to present to the Security Council at the suggestion of Nguefa, before the briefing by the head of MINUSMA El-Ghassim Wane who had not been consulted on the choice of this civil society briefer. Despite this, on March 23, 2023 the UN issued a human rights report indicating civilian deaths were up by 54 percent in 2022, and that at least 90 human rights violations were committed by FAMA, representing 26 percent of the total documented violence.

The positions of Russia and China in the Security Council comforted the Malian junta. France submitted a text calling for "thorough and independent investigations to establish the facts, find those responsible for these violations and abuses and bring them to justice" for approval by the Security Council on April 8, 2022, but Russia, backed by China, opposed it, deeming it "premature" as an investigation had supposedly already been opened by the Malian authorities. At the same time, the Russian Foreign Ministry congratulated Mali for an "important victory" against terrorism and described as "disinformation" the allegations of massacre of civilians by Malian forces and the involvement of Russian mercenaries in the operation. The Malian foreign minister continued to deny the presence of the *Wagner Group*—instead claiming it is all state-to-state support—and that Mali is "the victim of a war of disinformation aimed precisely at punishing the country for its choices."[80]

Russian Foreign Minister Sergey Lavrov's visit to Bamako on February 7, 2023, solidified the new security partnership, as deliveries Russian military equipment—including radar-guided surface-to-air missiles that the American civil aviation agency (FAA) expressed concerns about[81]—along with Turkish *Bayraktar* TB2 armed drones confirmed Mali was pursuing a military stabilization approach rather than the more political one advocated by the UN. *Wagner Group* mercenaries also continued to deploy further North in Mali, including in Ménaka on June 15, 2022, at the military base handed over to the Malian army by the French two days earlier.[82]

[79] Le Monde, "La junte malienne expulse le directeur de la division des droits de l'homme de la mission de l'ONU," February 5, 2023.

[80] Security Council, S/PV.9061, June 13, 2022.

[81] FAA Background Information Regarding U.S. Civil Aviation – Mali.

[82] Africanews, "Mali Wagner paramilitaries investissent la base de Ménaka", June 16, 2022.

At the same time, the Malian junta imposed increasingly strict restrictions on freedom of speech in Mali. In mid-March, following the broadcast of reports on atrocities committed against civilians by the Malian armed forces and *Wagner Group* in the context of clashes with jihadist groups, Bamako banned the two French media, RFI and France 24. Prior to this, a special envoy from the weekly magazine *Jeune Afrique* was arrested and expelled from Mali shortly after arriving in Bamako on 8 February, to investigate the private Russian company *Wagner*. As a result of this atmosphere, journalists and Malians in general became more cautious and often refused to talk. Opposition politician Oumar Mariko, who has been very critical of the Malian army, also went into hiding after representatives of the Malian security forces came to his home on April 4, 2022.[83] Malian authorities also stopped issuing visas to the UN Security Council Panel of Experts on Mali Sanctions.

In mid-May 2022, the Malian transitional government, which should have taken over the rotating presidency of the G5 Sahel (then held by Chad), announced that it was leaving the organization, including the Joint Force. This followed on logically from the divorce with France (and the European *Takuba* task force[84]), which had used its influence on certain G5 Sahel members—notably Chad and Niger—to put pressure on Bamako already under ECOWAS sanctions.[85] Ornella Moderan sees in this "a sovereigntist discourse that has gained momentum in Bamako in recent months, painting a picture of an anti-imperialist Mali that sets itself up as the scourge of a regional order presented as unjust and manipulated by the West, in particular France."[86]

Troop-contributing countries (TCCs) also started losing patience amid a series of disconcerting events. In June 2022, Mali's Foreign Minister strongly opposed the French offer to "continue its air support to

[83] David Baché, "Malil'opposant Oumar Mariko à nouveau inquiété par la justice," RFI, April 5, 2022.

[84] Morgane Le Cam and Elise Vincent, ""Mali Takuba" force a collateral victim of the junta's defiance of France" *Le Monde*, January 26, 2022.

[85] On July 3, 2022, ECOWAS decided to lift most of the sanctions imposed on Mali on the basis of a new timetable proposed by the junta for a transition period running until March 2024. However, ECOWAS firmly opposes the participation of the transitional authorities in the next presidential election, and maintains individual sanctions against members of the ruling junta, as well as administrative sanctions against Mali.

[86] Sophie Douce, "Mali's withdrawal signs de facto the death of the G5 Sahel", Le Monde, May 17, 2022.

MINUSMA" after the departure of the *Barkhane* force—support which had provided reassurance for many TCCs, as the UN mission lacked attack helicopters. On July 10, 2022, Malian authorities arrested 49 Ivorian soldiers—and describing them as "mercenaries" in an official communiqué[87]—who had come to support the German MINUSMA contingent upon their arrival at Bamako airport. The Malian junta tried to trade them in for Malian political figures in exile in Côte d'Ivoire[88] and held them for more than six months, until a Togolese mediation secured their release. The MINUSMA spokesperson was expelled later that month, accused of posting "unacceptable information" on Twitter about the case. After the arrests, Mali suspended all MINUSMA rotations for a month, until a "coordination meeting" was held to "facilitate the coordination and regulation of rotation" of UN contingents.

In this context, Sweden, the UK, Jordan, and Egypt withdrew their contingents, which made up the "mobile intervention force" based in Gao. Côte d'Ivoire started gradually withdrawing, as did Benin. On 14 July the Malian authorities this time prevented eight soldiers from the German contingent of MINUSMA from boarding at Bamako airport to leave the country[89] while Mali suspended all rotations of MINUSMA military and police personnel until a "coordination meeting" was held to "facilitate the coordination and regulation of the rotation" of UN contingents.[90] This prompted the German government to also consider the withdrawal of its troops and helicopters no later than May 2024. These cascading departures represented a quarter of the MINUSMA force and would not be easy for the UN to replace.

In this tense context and ahead of MINUSMA's June 2022 mandate renewal, UN Secretary-General Antonio Guterres himself expressed (again) serious doubts regarding the adequacy of MINUSMA:

> I'm not going to propose that we end this mission [MINUSMA], because I think the consequences would be terrible. But it's taking place in circumstances that really call not for a peacekeeping force, but a robust peace

[87] RFI, "Mali 49 Ivorian soldiers arrested in Bamako, "mercenaries" according to the authorities", July 11, 2022.

[88] RFI, "Militaires ivoiriens detenus au Mali: nouveau tournant dans la crise entre Bamako et Abidjan", September 12, 2022.

[89] Mahamadou Kane, "German Minusma soldiers prevented from leaving Mali", DW, July 15, 2022.

[90] RFI, "Mali suspends Minusma military rotations", July 15, 2022.

> enforcement and counter-terrorism force. This robust force must be an African force, of the African Union, but with a Security Council mandate under Chapter VII and with mandatory funding.[91]

In June 2022, the MINUSMA mandate was extended for another year but not without difficulties. The human rights issue was a particular sticking point, the P3 insisting on the need for peacekeepers freedom of movement to investigate allegations of human rights abuses, while several other Council members including Russia, Chine and African members felt that the text contained too many injunctions, infringing on the country's sovereignty. Mali's Ambassador to the UN Issa Konfourou conveyed the following message:

> the Government of Mali regrets that some of its strongest reservations were simply ignored in the final version of the resolution on which the Security Council voted. The Government of Mali reiterates its firm opposition to the freedom of movement of MINUSMA in the execution of its human rights mandate [...] Mali is not in a position to guarantee the freedom of movement for MINUSMA investigations without the prior agreement of the Government [...] The people of Mali expect MINUSMA to contribute to the search for lasting solutions to Mali's challenges. They do not want MINUSMA to become part of the problems in our country, and neither do they want human rights to be used for hidden political purposes. On behalf of the Government of Mali, I would like to thank Russia and China for abstaining from the voting on the resolution.[92]

It was indeed the first time a resolution on MINUSMA was not voted unanimously, as Russia and China both abstained, and for months before that already Council members had not been able to produce even a joint press statement on Mali. The final text of the resolution condemns "the use of mercenaries and violations of international humanitarian law and human rights abuses perpetrated by them" without mentioning *Wagner* by name, using the same wording as in resolution 2605 of November 12, 2021 on the Central African Republic.

The Council however expressed its support for the Secretary-General's proposal to conduct an internal review of MINUSMA within six months,

[91] TV5 Monde, "Mali: Antonio Guterres prône une force africaine sous mandat de l'ONU", May 6, 2022.

[92] Security Council S/PV.9082, June 29, 2022.

which it requested "be conveyed no later than 13 January 2023" in the form of a report which should include "a detailed analysis of the political and security challenges," "an assessment of the cooperation with the host authorities and movement restrictions," and make "recommendations on the necessary conditions for MINUSMA to continue operating" and give "options on MINUSMA's future configuration, force levels and ceiling of uniformed personnel."[93]

Statements made by Council members immediately after the adoption of the resolution were nevertheless very worrying, and suggest that there was no longer any common vision, if there ever was one. The French insisted that "the Malian transitional authorities must also shoulder their responsibilities," and the United States was alarmed by violations against civilians, including those committed by "terrorist armed groups, the Malian armed forces and the Kremlin-backed Wagner Group." The Russian representative, Anna Evstigneeva launched into a diatribe accusing the West of trying to manipulate MINUSMA:

> We are concerned that the resolution could be used by those who are displeased with the independent approach to foreign policy of the transitional Government to damage its reputation. We have become somewhat inured to the fact that our Western colleagues continue to feed the Security Council with false stories about Russia's allegedly destructive activities in Mali. They do not want to hear from us or our Malian colleagues, who have thanked Russia, along with others in this Chamber, for its bilateral assistance in combating terrorism. Moreover, we have not hear any complaints about us from other African partners. […] That raises the question of what their true motives are with regard to MINUSMA's activities. Despite the external pressures placed on them, we are confident that the United Nations and the Mission's leadership will remain impartial, act only in the interests of the Malian people and resolve all issues that arise through constructive engagement with the authorities of the host country.[94]

The UN Mission found itself under increased scrutiny, caught in the crossfire between Council members with divergent interests, knowing perfectly well that any action that could be interpreted as "not impartial" would expose it to accusations by one Council member or the other.

[93] Security Council Resolution 2640, June 29, 2022.

[94] Security Council S/PV.9082, June 29, 2022.

On the day of the departure of the last soldier from the *Barkhane* operation, on August 15, 2022, the Minister of Foreign Affairs of the Malian transitional government, Abdoulaye Diop, sent a letter addressed to the (rotating) President of the UN Security Council, Chinese Ambassador Zhang Jun, in which he accused France of repeatedly violating Mali's airspace, supplying arms and intelligence to terrorist groups, and spying on and intimidating the Malian armed forces.[95] Although Paris immediately denied these as "baseless" accusations, and no member of the Security Council effectively brought the Malian claims before the Council, this was yet another sign of the increasing attempts at instrumentalizing the UN.

In September 2022, Mali's interim Prime Minister Abdoulaye Maïga once again used the podium of the UN General Assembly to attack the UN as well as France and several of France's allied ECOWAS leaders, in a rather undiplomatic manner. He reiterated his government's "often-expressed" call for a "paradigm shift" and for MINUSMA to be "adapted to the environment in which it is deployed, and better coordinated with the Malian authorities." He also denounced "negative external influences" and the fact that his country had been "stabbed in the back" by the French authorities, whom he described as a "junta at the service of obscurantism" "nostalgic for neo-colonial, condescending, paternalistic and vengeful practices"; he also accused them of obstructing a debate in the Security Council on the issues raised in the letter his country sent to the President of the Security Council on August 15, 2022. Finally, he expressed his opposition to "the instrumentalization of the human rights issue for political purposes, positioning, even blackmail or intimidation."[96]

Following the request by the Security Council in its resolution 2640 (2022) for the UN to conduct an internal review of MINUSMA, the UN Secretary-General published a report in January 2023, in which he outlined three main options. The first option included an increase in the number of peacekeepers. However, the Malian authorities immediately indicated they are "not convinced of the relevance" of such an increase, which they had opposed in 2021. The second option was to keep the current strength of the mission but close bases in the Timbuktu and Kidal regions, redeploying peacekeepers to Ménaka and Ansongo. A third, more

[95] Morgane Le Cam and Christophe Châtelot, "La junte malienne accuse la France d'armer des groupes djihadistes, sans fournir de preuves", August 18, 2022.

[96] United Nations, "Mali's interim Prime Minister wants MINUSMA to adapt", September 24, 2022.

radical option was also considered: replacing the peacekeeping mission with a "political mission," which would have meant that there would be no more troops nor UN peacekeeping presence outside the capital Bamako.

These three options were not new and largely reflected the dilemmas the UN faced when designing MINUSMA in March 2013 and again during a 2018 independent strategic review, when, controversially, the United States disagreed with France by questioning the relevance of the peacekeeping model in the face of terrorism and pushing for an exit strategy for the mission. The 2023 context, however, was very different. The French counterterrorist "parallel force" had left Mali the previous summer. A new force of 1000 to 2000 men from the private military company *Wagner Group* and Russian instructors were now operating in Mali, mainly in the country's center. And Malian authorities had increased pressure on MINUSMA during 2022.

On 5 December 2022, during a meeting of the Under-Secretary-General for Peace Operations Jean-Pierre Lacroix and MINUSMA head El Ghassim Wane with the Minister for Foreign Affairs of Mali Abdoulaye Diop and other members of the Mali transitional government, the latter handed over a note articulating their priorities and expectations of the review. The note requested that the UN Mission "give top priority to the security dimension of its mandate" and to "support on logistics, fuel, food ration, medical evacuations, and intelligence to the FAMA [Malian armed forces]." They also demanded that the mission refrains from what they described as "politicizing and instrumentalizing the human rights issue." It also reiterated its demand that MINUSMA become "more offensive," something previous Malian governments had been asking for as well.[97]

The strategic choice made by El Ghassim Wane, whose first head of mission position it was and who had only been at the head of MINUSMA since March 2021, was to try to play down the difficult cooperation with the host authorities. In order to try to maintain the consent of the host country, he chose to insist on the good rapport he had with Foreign Minister Diop and "respect the sovereignty of Mali, which is free to choose its bilateral partners"[98]—but downplayed movement restrictions imposed

[97] See Annex to UN Secretary-General Report, Internal review of the United Nations Multidimensional Integrated Stabilization Mission in Mali - Report of the Secretary-General (S/2023/36), January 16, 2023.

[98] MINUSMA, Spokesman, Olivier Salgado, May 20, 2022.

on peacekeepers, which Bamako denied having taken, instead asking for "better coordination of MINUSMA's actions with the Malian State."

In the January 2023 report of the UN Secretary-General, the emphasis was on showing appeasement, emphasizing how peacekeepers participated in coordinated patrols with the Malian armed forced (81 during the previous quarter),[99] even though some UN staff including within MINUSMA did fear a "darfurization"[100] or a scenario *à la Eritrea*, where the peacekeeping missions experienced a decline "by a thousand cuts"[101] as host country consent continued to erode, or an "à la Carte"[102] UN Mission that would only focus on certain parts of its mandate but not others. At the same time, El Ghassim Wane also actively sought support from Council members, traveling to London, Paris, and Moscow at the end of 2022.[103]

Although French *Barkhane* troops had left Mali, tensions between Paris and Bamako persisted. France wanted to punish Mali. In January 2022, it has already pushed for the Council to express support for the sanctions decided by ECOWAS against Bamako following the junta's attempt to grant itself a five-year extension (to the 18-month transition period supposed to end with elections scheduled for February 27, 2022),[104] but Russia and China blocked the text proposed by France.[105] In November, France also suspended its development aid and pushed other Europeans and the World Bank to do the same.[106] For its part, Bamako accused

[99] UN Secretary-General Report, Internal review of the United Nations Multidimensional Integrated Stabilization Mission in Mali - Report of the Secretary-General (S/2023/36), January 16, 2023.

[100] The terms refer to the experience of the UN-AU hybrid peace operation in Darfur which faced serious limitations in its freedom of action by the Sudanese government of Omar al-Bashir, without the latter ever formally and completely withdrawing its consent to the presence of the mission.

[101] Julie Gregory and Lisa Sharland, Host-Country Consent in UN Peacekeeping, The Stimson Center, 2023, Washington, D.C., USA.

[102] Arthur Boutellis, "Will it Be MINUSMA à la Carte, or Geopolitical Endgame in Mali?", March 30, 2023.

[103] Africa Intelligence, "De New York à Moscou, dans le huis clos des négociations sur l'avenir de la Minusma," December 9, 2022.

[104] Using a recommendation from the Assises Nationales as a pretext. RFI, "Transition au MaliAssises recommend 6 months to 5 years of extension, without making a decision." December 30, 2021.

[105] "Mali Russia et Chine Bloquent À L'ONU Un Texte Supportant Les Sanctions De La Cedeao." *Le Point.*

[106] Morgane Le Cam, "Mali: la France suspend son aide au développement dans un contexte d'isolement croissant de Bamako," Le Monde, November 17, 2022.

France of violating its airspace and asked that France no longer hold the pen in the Council on the situation in Mali.[107]

There were reasons to believe Bamako would still prefer to keep MINUSMA for a bit longer while constraining its operations and ability to conduct human rights investigations in particular. MINUSMA's regular flights allowed Malian administrators and State to be present in the north of the country, the UN had increased its support to building and repairing infrastructure, including for the Malian military, and arguably the UN Mission had also "helped deter insurgents from taking over cities and larger towns,"[108] even if its record on protection of civilians and stabilization is mixed.[109] The securing of secondary towns and certain major roads by MINUSMA is certainly considered helpful by the FAMA and the Wagner Group, who can thus concentrate their efforts elsewhere, even if insecurity persists and attacks are getting closer to the capital Bamako.[110]

MINUSMA, which had its every action scrutinized, had no choice but to learn to live with a divided Security Council and a difficult host state. It tried to preserve the requisite impartiality needed to implement its ambiguous mandate crafted out of compromises between powers. It also became the target of "disinformation campaigns," and tried to walk a fine line between being drawn into an information war and remaining relevant, while also avoiding being seen only as a service provider for a military junta in power, or, worse, being an accomplice to crimes by turning a blind eye to certain actions by the host state. The mission was almost impossible. Even though MINUSMA continued to serve some of the divergent interests of the host state and permanent members of the Security Council, it risked becoming collateral damage.

[107] RFI, "ONU: Bamako ne veut plus de la France comme "plume" des résolutions sur le Mali au Conseil de sécurité," March 4, 2023.

[108] International Crisis Group, "La Minusma à la croisée des chemins," December 1, 2022.

[109] Sean Smith, "PROTÉGÉS OU DÉLAISSÉS: Les Civils s'Interrogent sur la Capacité de la MINUSMA à Protéger au Centre du Mali," Center for Civilians in Conflict (CIVIC), December 2022.

[110] RFI, "Mali: plusieurs morts après une double attaque près de Bamako," January 3, 2023.

The Surprise Mali Request and Rushed Departure of MINUSMA

The Security Council met in New York on June 16, 2023 to discuss the renewal of the MINUSMA mandate for another year. When briefing the Council, MINUSMA head El-Ghassim Wane insisted on:

> support for the ongoing transition process, the role we play as part of the international mediation … the Mission's presence in urban centres and its efforts aimed at protecting civilian populations; the multifaceted aid provided to initiatives by the Malian State to restore and extend its authority; the many projects carried out for the benefit of civilians in both the centre and the north and the contribution made to infrastructure development … it is impossible to overemphasize the fundamental importance of cooperation with the host country and the relationship of trust that has been built or should be built with it. It is helpful in that regard to recall the emphasis that the internal strategic review placed on the need for a stable and predictable relationship. There is a legitimate expectation on the part of the Malian people that our action will produce greater results, and the criticisms they have expressed are perfectly understandable. But only closer cooperation and a more trusting relationship can produce that greater effectiveness.[111]

As per usual, after every Council member spoke, the representative of the government of Mali was invited to speak and stunned the Council by calling for the "withdrawal, without delay" of the decade old United Nations Multidimensional Integrated Stabilization Mission in Mali. This sudden request which came as a surprise to the UN and most Council members and was apparently only decided a few days before, came after a long response to the UN report on the events in Moura (which had been published on May 12, 2023 and was followed by US sanctions against two Malian officers linked to the Moura massacre[112] as well as the *Wagner Group* head in Mali Ivan Alexander Maslov) denouncing an "attempt to instrumentalize or politicize the issue of human rights." Mali Foreign Minister Diop concluded:

> Options proposed by the Secretary-General in his report on the internal review of MINUSMA do not respond to the concerns and expectations

[111] Security Council S/PV.9350, June 16, 2023.

[112] Colonel Moustapha Sangaré and major Lassine Togola.

repeatedly expressed by the Government and people of Mali. We have no choice but to come to the objective conclusion that, given the track record of MINUSMA's ten-year-long presence, the Mission has failed to provide adequate responses to the security situation in Mali, and that the future outlook of its operations does not meet the security needs of the Malian people. Neither the Secretary-General's proposals, nor — and even less so — the draft resolution currently being negotiated by the members of the Council provide an appropriate response to Malians' expectations. The content of this draft resolution is so hostile to Mali that it justifies France's recusal as a penholder. Beyond security, which we are now ensuring through our own resources, we legitimately expected MINUSMA to be a mission that would assist and support the Government's efforts. Unfortunately, MINUSMA seems to have become part of the problem, fueling inter-community tensions exacerbated by extremely serious allegations that are highly detrimental to peace, reconciliation and national cohesion. The situation is creating a feeling of mistrust among the Malian population towards MINUSMA and causing a crisis of confidence between the Malian authorities and MINUSMA. In view of all of the aforementioned, the Government of Mali calls for the immediate withdrawal of MINUSMA. However, the Government is willing to cooperate with the United Nations in that regard.[113]

Although relations between the Malian transitional government and the UN mission had deteriorated, the timing of the demand still caught by surprise the UN Secretary-General's Special Representative and Head of MINUSMA El-Ghassim Wane, UN Secretariat, as well as most Council members. The Russian ambassador Nebenzia, who apparently was the only one to know in advance of the Malian request (the US had also reportedly received the information not long before the meeting), used the Council as an anti-Western tribune again:

Given the security vacuum in Mali following the inglorious withdrawal of the French Operation Barkhane and the European Union's Takuba Task Force, the Malian Government has been developing legitimate partnerships with international partners who can help to ensure security and protect the population. Thanks to support from Russia, among others, the Malian Armed Forces have been able to significantly increase their capacity to combat terrorism and convert it into real results on the ground ... The sharply negative reaction of some countries to the strengthening of the cooperation between Russia and Mali is a manifestation of neocolonialist attitudes and double standards. What needs to be done is to provide the Malian authorities with effec-

[113] Ibid.

> tive support rather than putting a spanner in the works, which is unfortunately what Paris and its allies continue to do, including by leveraging multilateral structures. From that point of view, the report of the Office of the High Commissioner for Human Rights published in May this year on the events in Moura in March 2022 looks like yet another politically motivated attempt to discredit Bamako's efforts to combat terrorism … One of the Malian Government's key tasks is combating terrorism, which does not come under the Blue Helmets' mandate. The fundamental element in the existing mandate is the protection of civilians and the provision of appropriate assistance to Mali in that regard. An excessive focus on internal political aspects and a shift to the protection of human rights in the peacekeepers' activities are undermining their core task of maintaining peace and security in the country.[114]

Some Council diplomats briefly considered options for keeping MINUSMA in place, but quickly the Security Council resigned itself to the fact that the host country's consent to the UN operation had been withdrawn and that the best that could be negotiated was a swift but orderly exit of MINUSMA's civilian and military personnel.

France as the penholder on Mali in the Security Council proposed a draft resolution that would "terminate MINUSMA's mandate as of June 30, 2023" but "maintain its personnel until 31 December 2023, to plan and execute the cessation of operations and transfer of tasks." But on 27 June, Mali and Russia, in what was apparently another concerted effort between the two, stunned the Council a second time when they opposed this already short six-month timeline for withdrawal (Russia blocking the approval of the MINUSMA budget in the UN General Assembly) and instead pushed for a departure of the mission within three months.

China—which had about 400 peacekeepers deployed as part of MINUSMA—and the African elected members of the Council (Ghana, Gabon and Mozambique at the time), played an important part in convincing Mali—and Russia—that six month was the bare minimal required for withdrawing the 13,000 peacekeepers-strong mission before liquidating its assets. The subsequent June 30, 2023 Council session on Mali was uneventful, showing a degree of resignation by Council member and notably the penholder France which did not even speak. The representative of the United States stated it "deeply regret the transition Government's decision to abandon the United Nations Multidimensional Integrated Stabilization Mission in Mali (MINUSMA) and the harm that that will

[114] Security Council S/PV.9350, June 16, 2023.

bring to the Malian people," but most other Council members simply acknowledged the Malian decision to withdraw its consent to the peacekeeping operation.

The withdrawal Resolution 2690 was adopted unanimously on June 30, 2023. The short two-page document decided the immediate termination of MINUSMA's mandate and "cessation of its operations, transfer of its tasks, as well as the orderly and safe drawdown and withdrawal of its personnel, with the objective of completing this process by December 31, 2023, in close consultation with the Transition Government of Mali,"[115] after which only few UN support staff would carry out the final liquidation of assets.

This meant that the planning for potential handover of MINUSMA tasks to the UN Country Team notably was rushed (even though the Security Council had asked MINUSMA multiple times before to start planning its exit strategy in previous years) and most Mission civilian staff had already left Mali by the end of September 2023. From August 2–4, 2023, a three-day workshop was organized by the UN to discuss the mission departure with the Malian transitional authorities of Mali. The latter stated their intent to assume responsibility for the political transition and support for the peace process, but were disappointed that no funding would be transferred to the government.[116] Bamako was already focused on its military offensive to regain territorial control of northern Mali, which would play a big part in maintaining its popular support.

The subsequent MINUSMA withdrawal proved challenging, not only because of the unprecedented short timeline for a peacekeeping operation of that size, but also because of the deterioration of relations between the Malian transition government—bolstered by *Wagner group* mercenaries and deliveries of new air assets and including Turkish *Bayraktar* TB2 armed drones—and signatory armed groups. At the end of December 2022 already, these groups had suspended their participation in the bodies implementing and monitoring the 2015 peace agreement and in January 2023 had withdrawn from the commission responsible for drafting the country's new constitution[117] denouncing the Malian transitional

[115] S/2023/611, Letter dated August 18, 2023 from the Secretary-General addressed to the President of the Security Council, August 21, 2023.

[116] The UN Secretary-General also issued a letter to the UN Security Council on the articulated plan of MINUSMA's withdrawal, particularly pertaining to the transfer of tasks.

[117] The constitutional referendum—an important step toward elections that would mark the end of the transition period—was initially planned for March 19, 2023, but was postponed earlier in the month.

government lack of commitment.[118] Prime Minister Choguel Maïga had previously called for an "intelligent and consensual rereading of the peace agreement" in which he saw the seeds of a partition of Mali, and in a letter to the International Mediation (which was leaked in March 2023), the Malian Minister of National Reconciliation accused the groups of "increasingly clear collusion with terrorist groups."[119]

The withdrawal of MINUSMA actually precipitated the resumption of fighting in northern Mali in the context of peacekeepers vacating their camps. Resolution 2690 (2023) requested that "MINUSMA team sites and assets are handed over in full compliance with general United Nations practices and financial regulations" and the Malian transitional authorities did not want these UN camps to fall in the hands of armed groups. Signatory armed groups (coalesced as Permanent Strategic Framework—CSP[120]), for their part, argued that the arrival of *Wagner Group* mercenaries and Malian armed forces attempts to regain territories in northern Mali—which they had been controlling, securing and exploiting (collecting taxes, notably on artisanal gold mining)—violated the 2015 peace agreement and previously signed ceasefires. The armed groups would not allow the army to take over the MINUSMA camps without prior negotiations.

On August 11, 2023, Malian armed forces and *Wagner Group* mercenaries clashed with CSP fighters as they vied for control of the Ber base, from which MINUSMA had not completed its withdrawal. The army took possession of the camp two days later, after the last MINUSMA personnel left. In September, the CSP—in some cases supported by non-signatory armed groups from the JNIM coalition of groups affiliated with al-Qaeda (which also imposed a blockade around Timbuktu in mid-August)—launched a series of deadly attacks on army outposts in northern and central Mali, notably in Léré in the Timbuktu region, Dioura in the Mopti region, and Bamba and Bourem, both in the Gao region in September, and Bamba and Tarkint in early October. In response, the Malian army and Russian allies carried out airstrikes on insurgent positions

[118] And calling for an urgent meeting of the international mediation outside Mali, a proposal that the authorities have thus far rejected.

[119] RFI, "Le Mali dénonce des violations de l'accord de paix par la Coordination des mouvements de l'Azawad," March 2, 2023.

[120] The Permanent Strategic Framework for Peace, Security and Development (CSP-PSD) includes the Coordination of Azawad Movements (CMA) along with the two factions of the Platform of the Movement of June 14, 2014 (Platform), and the Coordination of Inclusivity Movements (CMI).

and started labeling CSP attacks as “terrorist.”[121] The capture of the strategic town of Anéfis on 7 October, was a turning point, in that it put the rebel-stronghold of Kidal (about 110 km away) within TB2 drone range.

In MINUSMA’s departure mandate, the Council did not include any reporting requirements, but had requested that the government “fully respect all provisions of the Status of Forces Agreement (SOFA) until the departure of the final element of MINUSMA from Mali.”[122] But in an 18 August letter to the Council, the Secretary-General alerted that, “the timeline, scope and complexity of the withdrawal of MINUSMA are unprecedented.” The letter laid out some challenges, including Mali’s vast terrain, the presence of terrorist groups, the volume of the mission’s equipment, and recent developments following the 26 July coup in neighboring Niger—a key transit country both for the supply of MINUSMA during the drawdown and the exit of its personnel and equipment—and outlined withdrawal phases.[123]

By August 30, 2023, MINUSMA had only closed four of its smaller camps in Ménaka, Ogossagou, Ber, and Goundam. But the resumption of fighting and fear of getting caught in the crossfire forced the UN to rush the disengagement of its peacekeepers. From 24 September, Malian authorities blocked MINUSMA convoys in Gao from traveling to the Kidal region to retrieve equipment from the mission’s three camps in Aguelhok, Kidal city, and Tessalit, citing the security situation.

On 16 October, foreign minister Diop of the Mali Transition government issued a statement in which it expressed “astonishment” and “denounce[d] the unilateral action of MINUSMA” to withdraw from the Kidal advance posts which it had been operating with the reconstituted Malian Armed Forces Battalion (BATFAR), even though the UN had informed authorities well in advance of its plans to vacate or dismantle these outposts. The foreign ministry further criticized MINUSMA’s “passivity” as the CMA took control of the outposts and seized military and other equipment. The statement also indicated that Mali had suspended authorization of MINUSMA convoys to Kidal because of the security situation, which would be granted anew upon improvements in the situation, and that “Malian armed forces are continuing their deployment, with an

[121] International Crisis Group, “Northern Mali: A Conflict with No Victors,” October 13, 2023.

[122] UN Security Council, Resolution 2690 (2023), S/RES/2690 (2023), June 30, 2023, para. 4.

[123] Security Council, Resolution 2690 (2023), June 30, 2023.

aim of taking hold of MINUSMA camps, in accordance with UN rules and procedures."[124]

On 18 October, the UN warned that, "heightened tensions in Northern Mali increase the likelihood of the Mission being forced to depart without being able to retrieve equipment belonging to Troop-Contributing Countries or to the United Nations. This would result in significant financial losses."[125] and Under-Secretary-General for Peace Operations Jean-Pierre Lacroix and Under-Secretary-General for Operational Support Atul Khare took the initiative of briefing Council members on the matter on 19 October under "any other business." While some Council members considered the option of issuing a press statement on MINUSMA's withdrawal, Council dynamics continued to prevent any such statement from being issued.

On 19 October, a MINUSMA aircraft was hit by small arms fire while landing at Tessalit but there were no injuries to the crew or major damage to the aircraft. These incidents prompted MINUSMA to accelerate its departure from its camps, from Aguelhok on 23 October, Kidal on 31 October—which led CMA armed groups to temporarily take control of it[126]—then Mopti-Sévaré on 8 November, and Timbuktu by the end of December. The Special Representative of the Secretary-General El-Ghassim Wane, who had decided to stay "until the end," presided over a symbolic lowering of the flag ceremony organized in Bamako on 11 December[127] at which the Under-Secretary-General for Operational Support Atul Khare was also present.

MINUSMA eventually had to abandon and destroy lots of equipment for fear that it would end up in the hands of armed groups.[128] The UN Department of Operational Support estimated the cost associated with the forced abandonment of Contingent-owned and UN-owned equipment in Tessalit, Aguelhok and Kidal to approximately $83.5M, of which $57.9M is for potential reimbursement troop contributing countries. It took UN peacekeepers eight days to travel 350 kilometers from Kidal city to Gao in a convoy of more than 140 vehicles, traveling without air surveillance

[124] https://twitter.com/MaliMaeci/status/1714010471911551094

[125] Note to Correspondents: on Mali/MINUSMA, October 14, 2023.

[126] Reuters, "Mali rebels say they have taken base vacated by UN peacekeepers," 31 October 2023.

[127] https://minusma.unmissions.org/discours-du-chef-de-la-minusma-el-ghassim-wane-cérémonie-officielle-de-clôture-de-la-minusma-11

[128] In a 22 October note to correspondents, the UN announced that it had had to destroy equipment, such as vehicles, ammunition, generators, and other assets, before vacating the Tessalit base on 21 October. United Nations, "Note to Correspondents: MINUSMA withdrawal from northern bases", October 22, 2023.

because of the restrictions, during which time 37 peacekeepers were injured by improvised explosive devices.[129]

On November 14, 2023, Malian armed forces and *Wagner Group* mercenaries eventually entered the rebel stronghold of Kidal—and retook the MINUSMA camp there—without any fighting, but after having carried out a series of drone strikes in previous days which prompted both CMA fighters and a large part of the population to flee toward Mauritania and Algeria.[130] On 22 November, the Malian transition government nominated El Hadj Ag Gamou—a general in the Malian armed forces but who was also the leader of pro-Bamako Imghad Tuareg armed group GATIA—as the governor of Kidal.

On January 25, 2024, the Malian transition government declared the end of the 2015 Inter-Malian Agreement for Peace and Reconciliation that resulted from the Algiers Process and which implementation had been MINUSMA's main strategic priority and had been considered essential to stabilizing the country. The Malian authorities also launched a new "inter-Malian dialogue for peace and reconciliation"—which Transition president Assimi Goita had announced during its New Year speech—but which will not include CSP armed groups now considered as "terrorist" by Bamako.[131]

Assessing the Third Phase: Geopolitical Endgame

This third and final phase in the life of MINUSMA (2020–2023) was particularly eventful. It began with two successive military coups that put the planning for a MINUSMA exit strategy on hold indefinitely, followed by the arrival of *Wagner Group* Russian mercenaries in 2021 which precipitated the departure of the French "parallel force" *Barkhane*. The France-US divide that had marked the Council's dynamics in previous years was replaced by a more classic East-West divide, exacerbated by Russia's invasion in February 2022.

Initially, the Council managed to maintain a degree of political unity behind the two strategic priorities assigned to MINUSMA. The renewal

[129] United Nations Multidimensional Integrated Stabilization Mission in Mali, "MINUSMA closes its camp in Kidal, marking the end of its presence in the region", 31 October 2023. Also see Security Council Report, "UN Transitions in a Fractured Multilateral Environment," December 2023.

[130] Morgane Le Cam, "Au Mali, les drones de l'armée malienne frappent Kidal, bastion des rebelles du Nord", Le Monde, November 8, 2023.

[131] RFI, "Dialogue inter-malien: un comité de pilotage pléthorique mais avec des absences," February 2, 2024.

of the mandate in June 2020 was less controversial than previous ones, the result of an appeasement between Paris and the new American administration. The Franco-American rivalry over the succession to Mahamat Saleh Annadif as head of MINUSMA was also resolved after Washington conceded to France the appointment of El-Ghassim Wane.

Negotiations on the renewal of the mandate in June 2021 were more difficult for France. It first had to withdraw its proposal to increase the troop ceiling by 2000 blue helmets, and then saw this proposal (once reintroduced by the UN Secretary-General in mid-July 2021)[132] neutralized by the Malian transitional authorities in the context of discussions on an agreement with *Wagner Group* to send a thousand Russian mercenaries to Mali.

At the same time, the consent of the host state was eroding as support from Russia and *Wagner Group*'s mercenaries to the Malian armed forces grew. The Malian transitional government constrained the work of MINUSMA; it imposed no-fly zones for its aircraft, intimidated certain UN contingents, and denied MINUSMA access to the village of Moura to conduct a human rights investigation, despite pressure from several Western countries.

The positions of Russia and China—as well as that of India (non-permanent member) to a lesser degree—in the Security Council reinforced the Malian transitional government's position at a time when the Council was more divided than ever, as evidenced by its inability to agree on any joint statement to the press and Russia and China's first-time abstention on the June 2022 vote renewing MINUSMA's mandate.

The UN Mission, despite having tried to reassert a bit of its impartiality after the French "parallel force" departure from Mali and to make itself more useful to the Malian transitional authorities, paid the price for its—perceived or real—privileged relationship with France, a relationship that had often been more transactional than strategic. MINUSMA found itself caught between the penholder on Mali in the Council who seemed keen to punish the host state for having allowed in the *Wagner Group* and a growing Russian influence with it,[133] other permanent members with divergent positions (United States and China) or defiant positions in the case of

[132] Security Council, Report of the Secretary-General, "Adaptation and authorized strength of the United Nations Multidimensional Integrated Stabilization Mission in in Mali" S/2021/657, July 16, 2021.

[133] France, in addition to freezing French funding for several projects via Malian state structures, also put pressure on the EU and certain European partners, as well as the World Bank, to do the same.

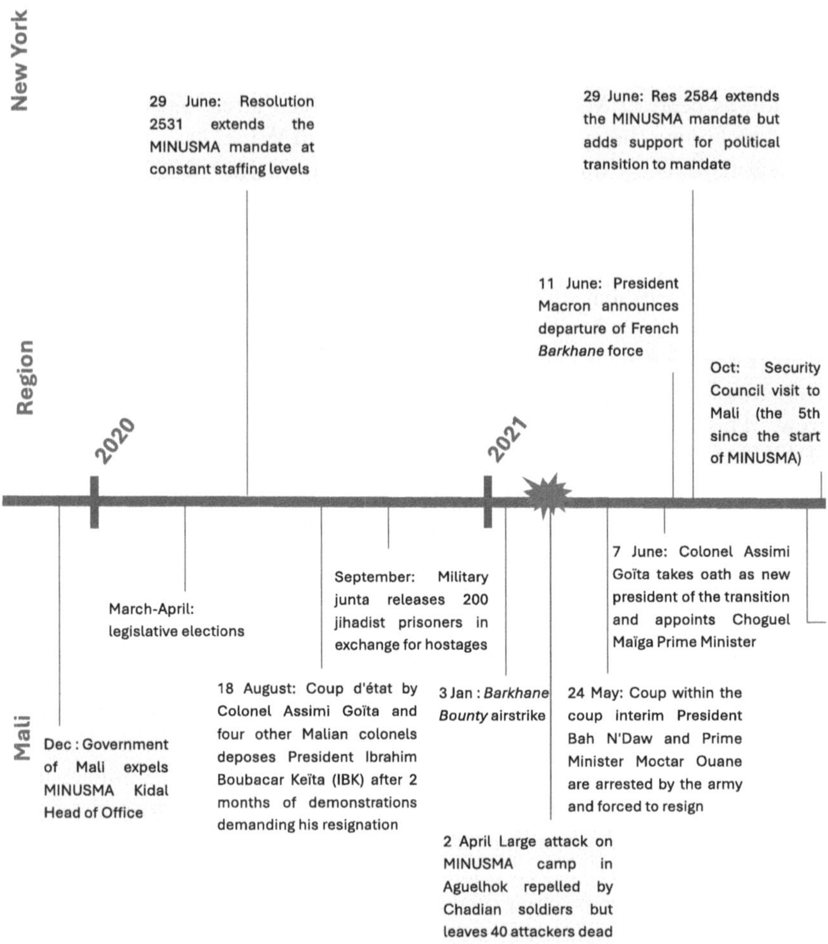

Fig. 4.1 MINUSMA timeline (2020–2024)

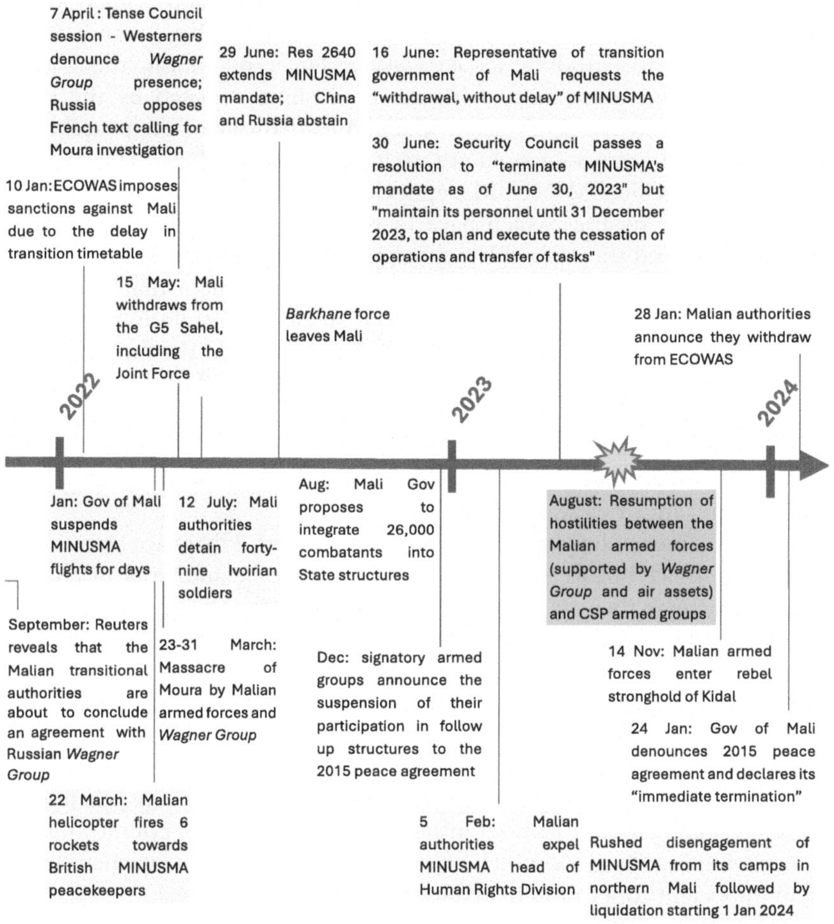

Fig. 4.2 (continued)

Russia, and a host state whose transitional authorities were seeking to consolidate their power at all costs, including by stirring up anti-French sentiment[134] and being more openly critical of MINUSMA. But unlike France, which decided to withdraw its *Barkhane* military operation, the UN—despite the many frustrations with the constraints imposed by the host country and the increasingly acknowledged inadequacy of the peacekeeping force to the context—would not itself decide to withdraw MINUSMA.

The dramatic final MINUSMA episode was a striking illustration of the evolving power rivalries which have increasingly played out around UN peacekeeping and their possible consequences on the multilateral system. The host country, represented by a transition government that came out of two successive military coups, and likely in close consultation or possibly even at the suggestion of Russia, called and obtained the "withdrawal, without delay" of the decade old United Nations Multidimensional Integrated Stabilization Mission in Mali. At this moment on June 30, 2023, as they unanimously voted the closing of MINUSMA, many Council members also sensed that the era of large, complex UN blue helmet missions in Africa was ending.[135]

In these times of heightened geopolitical rivalries, UN peace operations have little choice but to sometimes keep a low profile and work discreetly to try to preserve their relationship with the host state, so as not to become the collateral damage of Council divisions and power rivalries. Beyond the theatrical pronouncements that are part and parcel of multilateralism, allowing certain States to vent their frustrations on the multilateral stage, it is up to Council members and other member states to recognize the value of UN peace operations and not sacrifice peacekeeping on the altar of their power rivalries. These powers must be able to make use of peace operations while serving them, but without instrumentalizing them, or compromising or undermining them too much, as this would mean depriving themselves of an important tool of multilateralism in the future (Fig. 4.1).

[134] A major rally "to say no to the renewal of the Minusma contract" and "to support the operations of the FAMa [Malian armed forces]" was scheduled to take place in the Malian capital on 13 May, at the initiative of the Collectif pour la défense des militaires (CDM). The association, reputedly close to the ruling junta, had already spearheaded demonstrations against the presence of the French counterterrorist force Barkhane. Morgane Le Cam, "Au Mali l'avenir en pointilé des casques bleus," *Le Monde*, May 10, 2022.

[135] *International Crisis Group*, "What Future for UN Peacekeeping in Africa after Mali Shutters Its Mission?", July 10, 2023.

CHAPTER 5

Lessons from MINUSMA for UN Peace Operations in a Post-Liberal International Order

"A peacekeeping operation is not an army, or a counter-terrorist force, or a humanitarian agency. It is a tool to create the space for a nationally owned political solution."[1]

António Guterres, March 2018

Peace Operations and Counterterrorism: Constructive or Destructive Ambiguity?

Of the eleven countries most affected by terrorism in the world according to the *Global Terrorism Index,*[2] eight hosted UN peace operations in 2023, ranging from political missions (Libya, Yemen, Iraq, Afghanistan, Somalia, Colombia) to the most major peacekeeping operations (Mali and the Democratic Republic of Congo). The deployment of these operations in countries where there is little or no peace to maintain, and in "gray zones" where security threats are significant, adds to the complexity of the challenges facing the United Nations system, member states and their regional, national, and local partners.

[1] United Nations, "Peacekeeping: António Guterres calls for refocusing and strengthening operations - press releases." March 28, 2018.

[2] https://www.visionofhumanity.org/maps/global-terrorism-index/

A. Boutellis, *The UN Stabilization Mission in Mali (MINUSMA)*, Twenty-first Century Perspectives on War, Peace, and Human Conflict, https://doi.org/10.1007/978-3-031-66466-3_5

For a long time, UN peace operations had been protected from power rivalries because member states preferred to manage questions of national interest by themselves, including the fight against terrorism post September 11, 2001. However, UN peace operations have gradually, but inevitably, become the actors in these contexts, and the target of terrorist attacks. After the attack of August 19, 2003 on the United Nations headquarters in Baghdad which cost the lives of 22 people, the UN realized that the blue flag no longer protected its personnel. Among the victims was the UN Special Representative of the Secretary-General Sergio Vieira de Mello, who did not have succeed in the impossible mission of asserting the impartiality of the UN vis-à-vis the US superpower, which had just illegally invaded the country without a Security Council mandate.[3]

Over the years, peace operations have found themselves in partnerships and arrangements which brought them closer to parallel counterterrorist forces, at the risk of compromising their impartiality in the name of *realpolitik*. MINUSMA undoubtedly represents the most accomplished form of this "constructive ambiguity." Since 2016 its mandate under resolution 2295 asked peacekeepers "to anticipate, deter and counter threats, including asymmetric threats, and to take robust and active steps to protect civilians, including through active and effective patrolling in areas where civilians are at risk, and to prevent the return of armed elements to those areas, engaging in direct operations pursuant only to serious and credible threats."[4] MINUSMA was also mandated under resolution 2391 to provide operational and logistical support to the G5 Sahel Joint Force, a regional counterterrorism force which operated within a joint command mechanism with the French Operation *Barkhane* from 2020. In Somalia , the UN has also been supporting non-UN forces of the African Union Mission in Somalia (AMISOM and later ATMIS) whose main mission is the fight against terrorism.[5]

On the question of whether peace operations can themselves conduct counterterrorism operations, we easily agree with the conclusions of the 2015 HIPPO report which concluded that "owing to their composition and character, are not suited to engage in military counter terrorism

[3] United Nations, "Baghdad Attack, 15 Years Later: The UN Should Never Be a Target, Victims Emphasize | UN News." October 2018.

[4] Security Council, Resolution 2295, S/RES/2295 (2016), June 29, 2016.

[5] Walter Lotze and Paul D. Williams, "The Surge to Stabilize: Lessons for the UN from the AU's Experience in Somalia" New York: International Peace Institute, May 2016.

operations. They lack the specific equipment, intelligence, logistics, capabilities and specialized military preparation required, among other aspects. Such operations should be undertaken by the host Government or by a capable regional force or an ad hoc coalition authorized by the Security Council."[6]

But the issue is of course more complex, since the Security Council continues to mandate peace operations in contexts affected by terrorism. It is therefore imperative for these UN operations to adapt to these contexts even in the absence of a specific mandate to fight terrorism, given the human cost (asymmetric attacks caused more than 300 deaths among MINUSMA peacekeepers over a decade) and financial cost are high. For some political missions without peacekeepers operating in dangerous environments (Afghanistan, Iraq, Libya, Somalia, Syria, and Yemen), security expenses sometimes represent nearly 40 percent of their budget[7] even though their smaller footprints make them less exposed to human losses.

These peace operations have proven resilient, despite human losses among peacekeepers, particularly among more exposed African contingents. As explained previously, MINUSMA in particular has managed to adapt to these new threats thanks to training and equipment for contingents on counter-IED, the construction of bunkers in the camps and the installation of radars to detect mortar fire and other sensors.[8]

The challenge posed by the threat of terrorist attack nevertheless meant that MINUSMA had to devote most of its operational capacity to convoy escorts and self-protection. Between 2015 and 2017, the mission estimated that more than 80 percent of its peacekeepers were dedicated to protecting camps and convoys, and that very few peacekeepers therefore remained to implement the mission's civilian protection mandate. Despite the arrival of new units within MINUSMA after (dedicated combat convoy and demining units notably), the roads remained very dangerous, considerably limiting the movements of civilian personnel other than by air.

UN peace operations have always had to adapt to the changing nature of conflicts. This is partly why they became more multidimensional when dealing with internal (civil war) conflicts following the end of the Cold

[6] HIPPO Report, para 119 & 120.

[7] International Peace Institute (IPI), "UN Peace Operations in Violent and Asymmetric Threat Environments," March 2016.

[8] François Rihouay, "Above the UN camp in Kidal, a captive balloon under severe strain", Ouest France, February 27, 2016.

War. However, adaptation cannot be limited to managing the symptoms that are terrorist attacks, without looking at the causes of various insurgencies which are often linked to political, security and economic governance challenges in these countries.

Until today, peace operations operating in contexts affected by terrorism suffer from a lack of strategic coherence resulting from this "constructive ambiguity." Fruit of political compromises between Council members and the host government, this ambiguity often creates false expectations and misunderstandings. At the field level, this "ambiguity" has created recurring tensions between the MINUSMA civilian leadership of the mission—generally defending the basic principles of peacekeeping such as impartiality—and the military leadership (often Western in the case of the UN Mission in Mali) more inclined to support the Malian armed forces at all costs and to collaborate with the French counterterrorism "parallel force."

A central issue in such contexts is how the United Nations can preserve space for dialogue with all parties to the conflict. Security Council sanctions do not legally prevent the UN from speaking to armed groups including those designated as terrorists and their leaders. However, in practice, we observe a shrinking of the political space with growing pressure from certain powerful states and host states to delegitimize some of these armed groups and their leaders. Malian transitional authorities have notably used the "terrorist" label to discredit and delegitimize armed groups that were signatory to the 2015 peace agreement. But even previous Malian governments seemed keener on countering separatist armed groups with political agenda rather than the Al-Qaeda-affiliated JNIM coalition and ISGS-affiliated armed groups.

As John Karlsrud points out, the evolution of peacekeeping, " is likely to be counterproductive, as it will lead to more oppressive governments and more disillusioned people joining the ranks of opposition and terrorist groups, as well as undermine the UN in general and UN peace operations in particular."[9] Prior to the deployment of MINUSMA, the UN Secretary-General noted:

[9] John Karlsrud (2019) From Liberal Peacebuilding to Stabilization and Counterterrorism, International Peacekeeping, 26:1, 1–21, DOI: https://doi.org/10.1080/13533312.2018.1502040.

> it is critical that a clear distinction be maintained between the core peacekeeping tasks of an envisaged United Nations stabilization mission and the peace enforcement and counter-terrorism activities of the parallel force that will necessarily need to be established to preserve the hard-won security gains achieved so far. Any blurring of this distinction would place severe constraints on the ability of United Nations humanitarian, development and human rights personnel to safely do their work. If this were to happen, the United Nations would find it difficult to mount the kind of comprehensive system-wide response required to address the political, social and economic root causes of the multifaceted crisis in Mali.[10]

In 2017, the Departments of Peacekeeping Operations and Political Affairs even prepared an *Aide Memoire* on "Engagement with Non-State Armed Groups for Political Purposes: Considerations for UN Mediators and Missions" which noted:

> A number of UN initiatives have addressed the subject of dialogue, but none to date have been specifically devoted to the engagement of armed groups for political purposes.6 Such engagement – which includes a spectrum that extends from opening indirect channels to pursuing structured negotiations – may not always be desirable or even possible. Yet the "primacy of the political" in the UN's approach to fragile and conflict affected States suggests that engagement should always be considered.[11]

This *Aide Memoire* remained an internal UN document, but it clearly expresses the fact that it is essential that this political space be preserved so that peace operations can continue to carry out their work to help prevent, manage and resolve conflicts in an impartial manner.

As French forces left Mali in spring 2022, the question of possible negotiations between the Malian government and certain jihadist groups arose again, especially after one of the first decisions of the military junta when it came to power was to releases 200 jihadi prisoners in exchange for hostages in September 2020. The positioning of the UN vis-à-vis such groups and negotiations inevitably has implications for the perception of a UN mission, as was the case in Afghanistan at the time of the capture of

[10] Report of the Secretary-General on the situation in Mali, S/2013/189, March 26, 2013, paragraph 100.

[11] Aide Memoire "Engaging with Non State Armed Groups (NSAGs) for political purposes: Considerations for UN mediators and missions", internal UN document, May 2017.

Kabul by the Taliban.[12] Even though the UN published annual statistics of civilian victims of the conflict (including from operations by the Afghan army and the international coalition), UNAMA had the greatest difficulty preserving its impartiality and was still perceived as "Western" by the parties to the conflict.

Similarly in Mali, the perceived or real proximity between MINUSMA and the various non-UN "parallel forces" which carried out counterterrorism operations, including the Malian army, the G5 Sahel Joint Force and especially French *Barkhane* operation, has often been interpreted as UN peacekeeping sliding toward counterterrorism. This feeling was reinforced by the fact that peacekeepers were collocated with French forces in some camps, the fact that some of the same European troop-contributing countries operated under different forces and mandates in Mali (Denmark and the United Kingdom deployed both with *Barkhane* and MINUSMA[13]), and the fact that France held the pen at the Security Council on Mali. Jordan Street of the NGO *Saferworld* denounces the fact that permanent members of the Security Council "have proven unable to separate their own troubled investments in war-on-terror battlegrounds from their role as mandate creators for UN peace operations around the world."[14]

The Secretary-General's report of June 2018 also acknowledged that, "the proximity of MINUSMA, and its support role and cooperation with security actors, including counter-terrorism actors, had contributed to the perception that the Mission was engaging in counter-terrorism actions."[15] And yet, despite this and the pressure from the United States for "the Security Council to recognize that MINUSMA is not necessarily the best way to deal with terrorism in the region", France succeeded in maintaining the "constructive ambiguity" for the benefit of its national interest until the withdrawal of the *Barkhane* forces in 2022, even if it meant continuing to feed the perception of a MINUSMA which is neither

[12] Dulcie Leimbach, "Despite the Intensifying Dangers in Afghanistan, the UN Says It Is Staying Put," Passblue, August 14, 2021.

[13] Laurent Lagneau, "Barkhane/Mali Les Hélicoptères De Transport Lourd Danois Entament Leur Mission Sur Les Chapeaux De Roues - Zone Militaire", *Zone Militaire*, January 15, 2020.

[14] Ali Altiok and Jordan Street, "A fourth pillar for the United Nations? The rise of counter-terrorism," June 2020, p35.

[15] Report of the Secretary-General on the situation in Mali, S/2018/541, June 6, 2018, paragraph 59.

autonomous (necessitating a security guarantee including in extremis support from *Barkhane* notably) nor really impartial.[16]

A growing number of member states appear to subscribe to this approach, including China, which like Russia, tends to support the host state in the name of respect for sovereignty, even if it means turning a blind eye to human rights violations. In April 2022, China declared to the Council that "Mali is an important force in counter- terrorism operations in the Sahel region" and that while "China has noted the concerns in some media reports about human rights violations in the counter-terrorism operations in the Moura region ... China hopes that all sides will exercise restraint and refrain from making groundless accusations before any conclusion is reached by the investigation."[17]

In conclusion, the last twenty years teach us that peace operations are ultimately capable of operating in environments affected by terrorism, that this has a human and financial cost that the organization is able to assume to a certain extent. But the last decade also teaches us that the added value of the UN in helping address asymmetrical threats is not to deliver a decisive military response but to support and strengthen preventive, multistakeholder approaches to sustaining peace. The relevance of peace operations in these contexts will be judged on the latter rather than on the capacity to perfect an instrument to better manage the symptoms of asymmetrical threats so that peacekeepers can better secure themselves.

The experience of MINUSMA also leads us to recognize the limits and vulnerabilities inherent to major stabilization and peacekeeping operations more broadly. It must also be admitted that with the notable exception of Chadian soldiers, the majority of troop contributors are not prepared to assume the heavy losses that these operations entail. The majority of the 311 MINUSMA fatalities were among African (and to a lesser degree Asian) peacekeepers, with 80 fatalities for Chad only. Western contingents only had 12 fatalities overall.[18]

This therefore encourages us to consider alternative, more agile political models with smaller logistical footprints. Large multidimensional operations with thousands of peacekeepers, hundreds of civilian staff, gigantic "super-camps" and sizable aviation fleets, generally create expectations

[16] "France and its partners announce their military withdrawal from Mali." 2022. *France 24*.

[17] Security Council, S/PV.9012, April 7, 2022. p16.

[18] https://peacekeeping.un.org/sites/default/files/stats_by_nationality_mission_2_94_december_2023.pdf

from the part of the host state and population they can't meet and find themselves preoccupied with operational and security considerations which inevitably distract them from the essential "primacy of politics." The latter should remain the primary guiding principle and responsibility of both the UN Secretariat and all Member States, including the most powerful ones.

Competing Visions of the "Primacy of Politics"

The question of the "primacy of politics" in the conduct of UN peace operations is not new but it has been at the heart of debates on peace operations over the last twenty years. Already in 2000, the *Brahimi* report considered that the success of an operation largely depends on the "political support" it benefits from. The report insisted on the importance for the Security Council to give these operations "clear, credible and achievable mandates" so as not to repeat the fiascos of the 1990s in Angola, Somalia, Bosnia-Herzegovina and Rwanda.[19] And for good reason, peace operations are themselves political objects, as Jean-Marie Guéhenno points out:

> The greatest challenges to peacekeeping are not technical but political. This is not just because the most critical element for the success of a peacekeeping operation is the political process that must underpin it - and it is illusory to think that there is a technocratic solution to the challenge of state building - but also because the instrument of peacekeeping is itself a political enterprise, because different countries have different and often contradictory views on what they expect from the maintenance of peace and from the UN itself.[20]

Fifteen years later, the HIPPO report, while recognizing that "two decades of peacekeeping experience in more volatile settings calls for a flexible and progressive interpretation of those principles," also makes the "primacy of politics" the first of four major thematic axes of reform. The report notes that "Too often, mandates and missions are produced on the basis of templates instead of tailored to support situation specific political strategies, and technical and military approaches come at the expense of

[19] Report of the study group on United Nations peace operations known as the "Brahimi Report", A/55/305–S/2000/809, August 20, 2000.

[20] Guéhenno, *The fog of peace.* P292.

strengthened political efforts."[21] The 2019 Declaration of Common Commitments around the "Action for Peacekeeping (A4P)" initiative also made it the first of its priorities, formulated as follows: "to advance political solutions to conflict and enhance the political impact of peacekeeping."[22]

And yet, different, sometimes contradictory, visions persist on what the "primacy of politics" means and especially who, of New York or field operations, truly has primacy. As Richard Gowan notes , the A4P Declaration encompasses two quite different visions. The first, which would be "the diplomat's vision", places emphasis on the intergovernmental framework, and on the Security Council in particular which mandates and ensures regular monitoring of operations. According to the second—"the vision of the head of mission"—it is necessary to favor day-to-day management of political and operational dilemmas in the field,[23] and to concentrate on bringing together local, national and in certain cases regional actors around political solutions.

We could add a third intermediate vision which would be that of the United Nations Secretariat in New York. Indeed, the Secretariat often plays the role of transmission belt between the operation on the ground and the experts from the permanent missions of the Member States to the UN in New York, in particular between the regular briefings (generally quarterly) to the Security Council. The Secretariat also has a tendency toward self-censorship when renewing and adjusting mandates, generally only proposing options that it thinks are acceptable to the members of the Security Council, instead of "telling the Security Council what he needs to know rather than what he wants to hear" according to the famous recommendation of the *Brahimi report*. As we have seen in the case of MINUSMA, self-censorship also often leads heads of mission to play down tensions with the host country authorities and restrictions imposed on the work of peacekeepers.

Of course, these visions are not necessarily contradictory since the political strategies of UN peace operations must in principle take into account both the political-strategic realities of New York and the facts on the ground. As Adam Day and Jake Sherman point out:

[21] Ibid para 125.

[22] https://peacekeeping.un.org/sites/default/files/a4p-declaration-fr.pdf

[23] Richard Gowan, "The Politics of Action for Peacekeeping," UNU CPR, February 19, 2019.

> Overly prescriptive mandates can reduce the political space in which the SRSG, and by extension, the mission, operate. The political strategy—how the overarching political goals defined by the Council will be met—should be informed by an understanding of the drivers of conflict as well as possible points of influence, leverage, and opportunity. It should be driven by the SRSG, drawing on analysis of conflict drivers, including international, regional, national, and local dimensions, and input from a range of stakeholders both inside and outside the country. This does not preclude advice and guidance from UN Headquarters, which will often be more attuned to political dynamics in the Council and other international stakeholders than the field, but a political strategy should not be led from New York.[24]

But at a time when rivalries for power and influence between members of the P5 in particular are exacerbated, the first vision which is that of Council diplomats tends to gain the upper hand. This can sometimes put peace operations in difficult situations, either because Council members are not in unison in their political strategies; or because they impose unrealistic mandates (resulting from political compromises made following complex negotiations) on their operations, out of touch with the realities on the ground and the expectations of the host State and/or local populations. The addition of a second strategic priority in Central Mali to the MINUSMA mandate in June 2019, but with no increase in troop or financial resources to implement it, is a good example.

The proliferation of *strategic reviews* in recent years is also emblematic of certain powerful Member States attempting to take back control of the conduct of operations. Daniel Forti highlights the initial ambiguity of the first strategic reviews requested by the US administration of President Trump in particular. In the case of strategic reviews of special political missions in Afghanistan and Libya from 2017:

> Contentious debates over political strategy, both within these missions and between these missions and headquarters, informed the Secretariat's decision to appoint external leaders for these exercises. These were arguably the first independent reviews, though they were not conceptualized or referred to as such.[25]

[24] Jake Sherman & Adam Day, "Political Solutions Must Drive the Design and Implementation of Peace Operations," IPI Global Observatory, June 20, 2018.

[25] Daniel Forti, "Independent Reviews of UN Peace Operations: A Study of Politics and Practice," IPI, October 27, 2021.

It is indeed not insignificant that the use of this type of strategic reviews involving people outside the United Nations Secretariat became a common practice for a few years in a context of growing geopolitical rivalries, and a breakdown in trust between Council members, the UN Secretariat and field operations that different more or less powerful member states attempt to influence according to their respective strategic interests.

Resolution 2367 (2017) renewing the mandate of the special political mission in Iraq is particularly scathing in this regard since it requested the Secretary-General to carry out "an independent external evaluation of the structure and staffing of the Mission, the resources allocated to it, its priorities and areas where it enjoys comparative advantages and synergies with other United Nations entities, so that it and the United Nations country team are able to carry out tasks which have been prescribed to them in the most appropriate and efficient manner possible".[26] The United States , writing on UNAMI, even speaks of the need for an "impartial review by a third party," thus disavowing the United Nations Secretariat which is in principle itself impartial in its recommendations to the Security Council.

> These dynamics fueled an ongoing contest among council members, and between the council and the Secretariat, to influence the narrative on and oversight of UN peace operations. Council members (particularly penholder countries) have begun mandating more independent reviews not only to make findings more transparent but also to more concretely influence these strategic decision points.[27]

The 2018 MINUSMA strategic review was particularly damaging in many respects, dividing the members of the Security Council (France and the United States in particular) but also the mission on the ground which was suspicious of the intentions to the Secretariat in New York. The latter, under pressure from France in particular, ultimately did not make the strategic review report public, only adopting some of its recommendations in a selective and non-representative manner in the subsequent Secretary-general report on MINUSMA. The review did not prompt MINUSMA to review its political strategy either. The United States, on the other hand, wanted the entire report to be released, and denounced the fact that the

[26] Security Council, Resolution 2367, S/RES/2367 (2017), July 14, 2017.

[27] Daniel Forti, "Independent Reviews of UN Peace Operations: A Study of Politics and Practice," IPI, October 27, 2021.

mission had gradually moved away from its initial objective of maintaining peace, sliding toward supporting counterterrorist military forces.

The "stabilization turn" at the UN, as discussed in the introduction, has also had a significant impact on the "primacy of politics", even if stabilization reflects very different realities on the ground and has not been accompanied by a real doctrinal evolution.' Stabilization indeed carries inherent risks of "militarization" and "depoliticization" of peacekeeping approaches and issues. This is especially true in environments affected by terrorism. Peace operations also remain largely state-centered and dependent on local elites, despite the advent of the protection of civilians approach and discourses on (good) governance.

The majority of peacekeeping operations authorized over the past twenty years—from MINUSTAH to MINUSMA—have been "multidimensional" with mandates to help stabilize an area or country. This stabilizing action generally consists in helping restore the authority of the host State in a Weberian conception, and thus limiting the nuisance capacity of armed or criminal groups, by resorting to force if necessary. Despite the persistence of the "primacy of politics" in the discourse, the concern is that UN peacekeeping is drifting from "human security" toward "state security" or even in some case "regime security." Under the influence of certain permanent members of the Security Council, the UN sometimes endorses questionable elections in the name of a country's stability, as in the case of the 2018 presidential election in the Democratic Republic of Congo.[28]

John Karlsrud therefore expresses concerns that the concept of "liberal peace" is losing momentum, both at the conceptual level and in practice, due to the renewed interest of Western states in UN stabilization operations following long commitments in Afghanistan and in a context of financial crisis:

> the turn from liberal peacebuilding to stabilization and counterterrorism is likely to be counterproductive, as it will lead to more oppressive governments and more disillusioned people joining the ranks of opposition and terrorist groups, as well as undermine the UN in general and UN peace operations in particular.[29]

[28] United Nations, "DRC: UN chief takes note of announcement of provisional results of presidential election - press releases." January 10, 2019.

[29] Karlsrud, "From Liberal Peacebuilding to Stabilization and Counterterrorism."

While it is sometimes tempting to blame power rivalries for the failures of peace operations, this should not elude the fact that even in certain contexts where the Security Council is united behind a common strategy, such as in South Sudan, or in the Central African Republic and Mali (before the arrival of *Wagner Group* Russian mercenaries), peace efforts have not always succeeded. And UN operations must adapt the way they work accordingly. This is also what Jean-Marie Guéhenno suggests:

> The UN must focus on politics, using its comparative advantage – its independence – winning the trust of the parties, while preserving its access to the major powers to put pressure on them. However, it is challenged by growing divisions within the Security Council, the changing nature of conflicts and a crisis of states that reflects long-term trends. This is no reason for the UN to abandon its role in ending civil wars, but it must recalibrate its ambitions and adapt its approach: to be less state-centric and more inclusive; more robust militarily; and more disciplined in its priorities.[30]

Indeed, the increasingly fragmented nature of conflicts and armed groups must also lead to a rethinking of conflict resolution approaches, including at the local level (with more "bottom-up" approaches) in order to better take into account the needs of communities. Peace operations can help link local peace initiatives—which are useful but often not sustainable in the medium term—with more formal national peace processes. Only the latter can provide more lasting solutions to conflicts that are increasingly often regionalized or even internationalized, but which find their origins in growing inequalities, a failing social contract due to the gap between the ruling elites and the aspirations of their people, and states institutions that are not representative of the populations of a country.[31]

Approaches to peacekeeping traditionally centered on the host state and the strengthening of its capacities and its "authority" are indeed increasingly called into question today, all the more so when the states in question lack legitimacy. In Mali for example, the mandate to support the return or "extension of State authority" to the North was initially part of a stabilization framework and it was only after 2015 that it became part of

[30] Jean-Marie Guéhenno, "The United Nations and Civil Wars", *Daedalus* 147, no [1] (January 2018): 185–96, https://doi.org/10.1162/DAED_a_00483.

[31] Arthur Boutellis, Delphine Mechoulan, and Marie-Joëlle Zahar, "Parallel Tracks or Connected Pieces? UN Peace Operations, Local Mediation, and Peace Processes," International Peace Institute, December 2020.

the implementation of the Agreement for Peace and Reconciliation. Such approaches are particularly problematic in contexts where the State has had little or no historical presence and/or was not accepted in certain parts of a country.[32] This underlines the difficulty or even the impossibility of a stabilization process solely based on *state-building,* and the fact that a State is never built from the outside.

Finally, the vision of the "primacy of politics" in peace operations also increasingly comes up against the fact that in many contexts, the UN is no longer necessarily the primus inter pares, and becomes one actor among others. In the most extreme cases, the finds itself relayed to the role of powerless observer and service provider—the UN mission in Mali had sometimes been nicknamed "Air MINUSMA." The tendency of peace operations to carry out more and more projects financed by donors through trust funds can also distract the organization from its primary purpose if these logistical and programmatic capacities are not used and leveraged in support of advancing political solutions.

In most cases, the UN has lost its role as "impartial chief mediator" in conflicts and often finds itself in the position of deputy of a regional state which has assumed the role of mediator. This was the case in Mali when Algeria established itself as the international "lead mediator", a role it assigned itself after the failure MINUSMA and ECOWAS to implement the June 2013 Ouagadougou interim agreement and the May 2014 Kidal clashes, but a role it also retained in the implementation phase of the 2015 agreement despite MINUSMA's significant presence of 13,000 MINUSMA peacekeepers and civilian components on the ground in Mali.

Several questions arise: should United Nations peace operations resign themselves to being just one mediator and political player among others with more or less influence depending on the context and the state of power rivalries and space great powers and regional powers are willing to grant the UN? Or should UN peace operations on the ground continue to try to play the role of leader or conductor rather than risk finding themselves in a situation in which it supports regional or sub-regional organizations themselves without or under influence depending on the contexts (the limited role played by the sub-regional East African organization IGAD in the Sudan and Ethiopia conflicts is a case in point)?

[32] Due to the fact that not only was it not accompanied by services to the populations, but was often accompanied by abuses by defense and security forces perceived as not being legitimate since they were not representative of local communities.

Or can these peace operations reinvent themselves and become less "state-centered" and find a more impartial position between the belligerents and the different, sometimes contradictory interests of big and middle powers? Can they better take into account the aspiration of people for a new "social contract" in contexts of political transition such as in Sudan? Can they manage the significant countercurrents from Member States in New York who have continually challenged the promotion of *people-centered approaches* by parts of the United Nations Secretariat and the "Third UN" (non-governmental organizations and other *think tanks*)?

Pushback Against "Liberal" Human Rights and PoC Norms

The last twenty years have been marked by the expansion of liberal norms within UN peace operations and in particular the generalization of protection of civilians (PoC) mandates—first introduced in the mandates of the missions in Sierra Leone and the DRC—for large multidimensional peacekeeping operations. This trend has been accompanied by a real acculturation among Member States, but also by a growing integration of the question of human rights with dedicated components in peace operations. The Office of the High Commissioner for Human Rights (OHCHR), based in Geneva, even opened a branch dedicated to peace operations in New York in 2015—the Support Section for Peace Missions. Since 1999 and following the tragedies in Rwanda and the former Yugoslavia, 14 peace operations have been mandated to protect civilians from physical violence. As of 2021, eight of the fourteen active United Nations peacekeeping operations have a PoC mandate.

Following the relative successes of human rights missions led by the Department of Political Affairs in El Salvador (1991–95) and Guatemala (1997), human rights were gradually integrated into certain peacekeeping operations (Abkhazia and Eastern Slavonia) in 1996, then Angola in 1997 and Kosovo and East Timor in 1999. In 2023, twelve peace operations include human rights in their mandate, and eleven of them have within them a "human rights" component dedicated to the implementation of this mandate in cooperation with the other civilian components of the mission as well as the peacekeepers and the UN police.[33]

[33] Céline Giuliani, "Advancing Peace through Human Rights in UN Peace Operations: A Policy Perspective on Achievements and the Way Forward," *Journal of International Peacekeeping* 24, no 1–2 (May 24, 2021): 252–84, https://doi.org/10.1163/18754112-20210006.

However, in the current context of growing power rivalries but also of affirmation of host countries, we observe a clear decline in human rights in part due to the readiness of Security Council members to compromise on UN values in the name of *realpolitik*. In the case of MINUSMA, the issue of human rights became one of the major points of contention between the UN mission and the host country government (and its Russian ally).

But this is not unique to Mali. The latest political mission authorized by the Security Council in 2020, the UN Integrated Transition Assistance Mission in Sudan (UNITAMS) is another example. The negotiations of Resolution 2524 highlighted significant divisions in the Security Council on issues relating to human rights and the protection of civilians. Russia and China but also the three elected African members (A3) opposed the inclusion of PoC tasks in the mandate[34] which means UNITAMS could only support host government's actions to protect civilians but had no direct responsibility for it.[35]

Even if the question of the protection of civilians enjoys "moral unanimity" in the context of peacekeeping operations, "pro-sovereignty" states such as Russia and China but also a certain number of powerful emerging countries such as Brazil or India, take a dim view of a PoC which could be carried out in spite or even against the host state in countries like Sudan and South Sudan, or justify military interventions as in Libya in 2011.

More than twenty years ago, the *Brahimi report* of 2000 suggested strengthening the UN's capacities to develop peacebuilding strategies and to deploy multidisciplinary civilian teams, including human rights experts. It also suggested strengthening the capacity of the Office of the United Nations High Commissioner for Human Rights (OHCHR) to support the "human rights" components within UN peace operations, a practice which had already begun in certain missions in the 1990s.

The issue of human rights then gradually became institutionalized in the peacekeeping and political affairs departments. A first Memorandum was signed in 2002 between the OHCHR and DPKO, formalizing the participation of the OHCHR in the planning of peace operations as well

[34] Security Council Report, "Sudan: Briefing on the UN Integrated Transition Assistance Mission in Sudan and the AU/UN Hybrid Operation in Darfur," What's In Blue, December 7, 2020.

[35] https://unitams.unmissions.org/en/support-civilian-protection

as the famous "dual reporting" system. The head of the human rights component is thus the human rights advisor to the head of mission and the representative of the United Nations High Commissioner for Human Rights within the mission, which means that he/she reports both to the head of the peace mission but also directly to OHCHR in Geneva to avoid "political censorship," without completely preventing it in practice since we often see reports on violations human rights deemed "sensitive" which publication gets postponed.

A Secretary-General's Directive was then published in 2005 on "human rights in integrated missions," requesting that human rights be fully integrated into peacekeeping operations but also into special political missions, and into the selection and the evaluation of heads of peace operations. It also calls for a regular report dedicated to human rights in the country to be made public independently of regular reports to the Security Council, and for the UN Secretariat to request mandates from the Council that address human rights issues.

In 2011, OHCHR, DPKO, DPA and the Department of Field Support (DFS) finally adopted a landmark policy on the integration of human rights into peacekeeping operations and political missions. Member States also accepted the principle of financing the "human rights" components of peace operations with the peacekeeping assessed budget (and not the OHCHR budget alone, the latter being financed by the regular budget of the organization). This allowed a considerable increase in the personnel of the human rights components of peace operations on the ground.

The past decade has also seen the adoption of important human rights policies that directly impacted the work of peace operations. Based on the experience of the UN operation in the DRC and its "conditionality" policy in its support to military operations of the national army, the UN developed in 2011 a human rights due diligence policy (HRDDP) for contexts where peacekeepers provide support to non-UN security forces. The policy was made public in 2013.[36]

In 2012, a policy of background checks for respect for human rights (the *Screening policy*) for contingents serving under blue helmets was also adopted. It allows the UN to refuse or in rare cases to repatriate certain

[36] Jérémie Labbé and Arthur Boutellis, "Proxy peacekeeping operations: consequences of UN peacekeeping partnerships with non-UN security forces on humanitarian action", International Review of the Red Cross, January 2016.

contingents.[37] The 2013 *Human Rights Up Front initiative* was another effort by Secretary-General Ban Ki Moon to promote respect for human rights at the end of his second term, although this issue did not feature prominently in the 2015 HIPPO report.[38]

His successor, Antonio Guterres has been criticized for his *realpolitik*. He seemed to consider that it is "counterproductive" to risk alienating great powers and has been cautious with leaders who openly display their disdain for human rights. These include Donald Trump, Vladimir Putin, and Xi Jinping, leaders of permanent members of the Security Council; but also Recep Tayyip Erdogan of Turkey, Mohammed Ben Salman of Saudi Arabia,[39] or Narendra Modi of India.[40]

The UN High Commissioner for Human Rights, Zeid Ra'ad Al Hussein, announced at the end of 2017 that he was not applying for a second four-year term due to the "geopolitical context" which "might involve bending a knee in supplication; muting a statement of advocacy" and diminishing "the independence and integrity of my voice."[41] Andrew Gilmour, Antonio Guterres's former human rights aide who left the United Nations in December 2019, also denounced the secretariat's excessive deference to powerful member states in an email obtained by *Foreign Policy*:

> the impression now conveyed by some is that standing up for [U.N.] Charter values and speaking out for principles over expediency, and on major violations of human rights and international law, is somehow a bit quaint, even a sign of zealotry and political imbecility.[42]

And *Foreign Policy* concludes, "several U.N. officials believed that the growing resistance to the rights initiative in Guterres's office marked an

[37] RFI, "Central Africa: why the UN sends back the Gabonese peacekeepers of Minusca", September 15, 2021.

[38] "UN Deputy Secretary-General presents "Rights First" initiative", December 19, 2013.

[39] Saudi Arabia was removed from a "black list" violating the rights of children in armed conflict in 2016. Catherine Gouëset, "How Saudi Arabia was removed from the country's 'list of shame' UN", L'Express, June 10, 2016.

[40] Marie Bourreau, "The UN Secretary-General accused of forgetting human rights", Le Monde, February 25, 2020.

[41] France 24, "The UN High Commissioner for Human Rights will not seek a second term." December 20, 2017.

[42] Colum Lynch, "UN Chief Faces Internal Criticism Over Human Rights." Foreign Policy. February 4, 2020.

early act of surrender to big-power politics and the forces of nationalism and populism sweeping across the globe, with a new generation of autocratic leaders less willing to brook U.N. criticism of its treatment of its citizens."

These criticisms undoubtedly motivated Secretary-General Antonio Guterres's "call to action for human rights" of 2020 at the time of the 75th anniversary of the organization. This appeal underlined that "human rights are the responsibility of each actor in the United Nations system and that the culture of these rights must be anchored in all our activities, both in the field and at the regional level and at Headquarters." And peace operations, among others, must base "the execution of their mandate and their action on an analysis of risks and prospects in terms of human rights which takes into account, among other things, gender issues."[43]

The reforms carried out by Antonio Guterres in 2019 also consolidated the protection functions of peace operations (sexual and gender-based violence, children in armed conflict, PoC) within the "human rights" component of these operations, for consistency and increased efficiency. This was done in MINUSMA and MINUSCA, and the most recent political mission, UNITAMS in Sudan, also integrated protection and rule of law functions within the same structure, which is not without creating friction with the host State.

Yet neither the Secretary-General's appeal, nor the internal restructuring of the United Nations Secretariat in New York, nor even the HRDDP, have been convincing. The reason is that human rights are subject to "repeated assaults from the great powers: from the United States against sexual and reproductive rights or against the International Criminal Court, from China and Russia which muzzle civil society and obstruct everything that is linked to human rights, whether in political resolutions, in the UN organization chart or in its budgets."[44]

Symbolically, the fact that the "third UN" of representatives of civil society organizations, NGOs and *think tanks* was not authorized to access the United Nations Secretariat building in New York for almost two years under the pretext of the COVID-19 crisis (while diplomats and journalists were allowed in the building during the General Assembly in September

[43] United Nations, The Highest Aspiration: A Call to Action for Human Rights, 2020.

[44] Marie Bourreau, "The UN Secretary-General accused of forgetting human rights", Le Monde, February 25, 2020.

2021) also reinforced the idea that the current Secretary-General practices a double standard.[45]

Céline Giuliani, a former UN employee, explains the difficulty of the work of UN human rights experts in peace operations:

> Peace operations are facing widespread human rights violations and abuses against civilian populations, while the consequences of the climate crisis and the global Covid-19 pandemic are exacerbating both conflicts and inequalities. Multidimensional peacekeeping operations with civilian protection mandates have the impossible task of protecting civilians in contexts where there is little or no peace to maintain, where host state institutions are weak, and where no one is not held responsible for violations, corruption and lack of public services.[46]

At the same time, she insists on the deleterious effect that power rivalries between Member States have had over the last decade on this work of defending human rights on the ground, "with the central nature of human rights for the peace and security and commitment to multilateralism increasingly contested." An additional risk for a certain number of peace operations is that of being complicit in human rights violations committed by the national or international defense and security forces they support, from Mali to Somalia and the DRC.

Peace operations however have not been as much victim of the attacks of certain great powers as was the Human Rights Council in Geneva. The human rights mandates of peace operations do not currently appear to be threatened as such, and some have even been strengthened in recent years, from Somalia to Libya and Afghanistan, as well as Mali and Sudan.

In 2020, the United Nations High Commissioner for Human Rights was even invited by France to brief the Security Council on the situation in the DRC and by Germany for the first thematic debate on "peace operations and human rights" in July 2020. Michelle Bachelet, who did not hesitate to criticize the French government the previous year for its "violent use and excessive force" during the Yellow Vest demonstrations,[47]

[45] Colum Lynch and Robbie Gramer, "NGOs Frozen Out of UN Building." Foreign Policy. September 23, 2021.

[46] Giuliani, C., 'Advancing Peace Through Human Rights in UN Peace Operations: A Policy Perspective on Achievements and the Way Forward', 2020.

[47] Badie, *France , a thwarted power*. P228–229.

ended her intervention with a request for support from the Security Council:

> UN peace operations are among the Organization's most important achievements and a powerful tool for protecting and promoting human rights. They need the resources, and your strong political support, to bring all UN operations together around an effective common approach to crises, from prevention to recovery.[48]

Shortly after, Michelle Bachelet herself became a victim of power rivalries. Pushed by the United States which had criticized her for her discretion on violations of the rights of Uighurs in the Xinjiang region, she finally made an official visit to China in May 2022, the first by a UN high commissioner for human rights since 2005. The visit was tightly controlled by Beijing and the report of the visit was slow to come out. Michelle Bachelet immediately announced her decision, like her predecessor, not to seek a second term.[49] The report, which would be made public just 15 minutes before Michelle Bachelet left office, concludes that "serious human rights violations" are being committed against Uyghurs and "other predominantly Muslim communities."[50] The spokesperson for the Chinese Ministry of Foreign Affairs described the UN report as "a patchwork of disinformation and a political tool at the service of the United States and certain Western forces to contain China using the subject of Xinjiang."[51]

Divisions within the Security Council between the Western P3 and the P2 seem to be growing. On the one hand, the P3 (France, United Kingdom and United States) and certain elected members (notably the Europeans), informally consult human rights experts when formulating mandates. They regularly raise the question of the human rights situation in open and closed debates on peace operations. Every year, they organize

[48] United Nations, "Security Council begins reflection on "peace operations and human rights" – press releases." July 7, 2020.

[49] Serge Enderlin, "Michelle Bachelet will leave her post at the United Nations", Le Monde, June 14, 2022.

[50] UN, "China commits "serious human rights violations" in Xinjiang according to OHCHR report," August 31, 2022.

[51] Global Times, "China denounces OHCHR report on Xinjiang", September 1, 2022.

an informal meeting using the *Arria formula*, to which the heads of the human rights components of the various peace operations are invited.[52]

On the other hand, the P2 (Russia and China) and many countries of the so-called global south or G77 view the human rights agenda applied to peace operations with suspicion, in the name of respect for state sovereignty. For example, these countries supported Colombia's demand not to have a "human rights" component within the peace operation on its soil and blocked the prioritization of human rights during the mandate renewals for MINUSCA mandate[53] and for MINUSMA in June 2022 on the pretext that this would undermine the sovereignty of the country. Chinese diplomats have even tried in recent years, but without success, to reduce the budgets allocated by the Fifth Committee to the "human rights" components of peace operations.[54]

What hasn't help is the fact that over the past decade or so, many injunctions from Western powers have been perceived as attacks on the sovereignty of States. In 2011, China, Russia, and other E10 member in the Council openly criticized the P3 for abusing liberal norms to justify military interventions and regime changes. They refer in particular to Côte d'Ivoire in the context of the post-electoral crisis (which led to the arrest of President Gbagbo[55]), but especially to Libya where the NATO intervention precipitated the end of Qaddafi's regime and ultimately contributed to the lasting destabilization of the Sahel.[56]

Likewise, the discredit that France attempted to cast on the report of the MINUSMA human rights division on the Operation *Barkhane airstrike* near the village of Bounty in Mali on January 3, 2021[57] contributed to weakening the human rights agenda in peace operations and the impartiality on which it is based. Moreover, UN peace operations often self-censor themselves when human rights violations concern a member of the

[52] 'Brief on Arria Formula meetings, Security Council Report, UN Security Council Working Methods,' December 16, 2020.

[53] Security Council Report, "Central African Republic: MINUSCA Mandate Renewal*," What's In Blue, November 11, 2021.

[54] Louis Charbonneau, "China Pushes to Cut UN Human Rights Posts," Human Rights Watch, June 7, 2017.

[55] Arthur Boutellis, "The Security Sector in Côte d'Ivoire: A Source of Conflict and a Key to Peace," New York: International Peace Institute, May 2011.

[56] Libération, "Nato ignored the presence of Gaddafi in the attacked convoy," October 21, 2011.

[57] United Nations, Report on the Bounty incident of January 3, 2021, March 2021.

P5, as in the case of sexual abuse of minors attributed to around fifteen of French soldiers from *Sangaris* Operation in the Central African Republic, which the MINUSCA hierarchy allegedly tried to stifle.[58]

Roland Paris rightly notes:

> Many scholars have pointed out that peace missions since the end of the Cold War have tended to promote a particular kind of state—liberal market democracy—reflecting the pre-eminence of the US and other Western countries (and of their political and economic models and ideologies) in the international system since the collapse of the Soviet bloc. Post-Cold War peace operations, in this sense, have 'enacted' the prevailing constitutive norms of the world political system. However, the norms and understandings that underpin liberal peacebuilding will likely be challenged as rising actors, who do not necessarily share these assumptions, gain greater influence and voice in international affairs.[59]

Other scholars "argue that the value underlying liberal peace is not universal or causally linked to peace but reflects the ideological preferences of the world's most powerful actors."[60]

At the end of the two years 2019–2020 that Germany spent as a non-permanent member of the Security Council, and during which it was a champion of human rights, Ambassador Christoph Heusgen called on China to release two Canadian detainees for Christmas 2020. The Russian representative then replied to the German representative, "we will not miss you" and the Chinese representative added, "I would like to say something from the bottom of my heart: good riddance."[61]

It remains to be seen whether Germany's legacy on human rights will be perpetuated by the next E10 members and whether the Western P3 will be robust enough, and whether the UN bureaucracy and peace operations on the ground will be resilient enough, to counter the Chinese steamroller in the multilateral system and Russia's policy of aggressive influence in Africa in particular. These models of authoritarian governance resonate with certain G77 member states, particularly in these times of crisis of the

[58] RFI, "RCA: French soldiers suspected of sexual abuse of minors", April 29, 2015.

[59] Roland Paris, "The Geopolitics of Peace Operations: A Research Agenda," *International Peacekeeping* 21, no 4 (August 8, 2014): 501–8, https://doi.org/10.1080/13533312.2014.946743. P504.

[60] Barkawi and Laffey, "The Imperial Peace."

[61] Nichols, Michelle. "'Good Riddance,' China Says as Germany Leaves UN Security Council." Reuters. Thomson Reuters, December 22, 2020.

post-Afghanistan model of liberal democracy,[62] and benefit from the resurgence of a spirit of non-alignment or multi-alignment of countries from the global south—including those ruled by military juntas arrived in power through coups— in these times of new Cold War.

The Experience of the Limits of "Integrated" and "Multidimensional" Approaches

UN peace operations have become a unique model of an "integrated approach" to stabilization over the past two decades. They associate politics, embodied by the head of mission who is often a civilian with the profile of a former politician or diplomat, the military with peacekeepers, and the so-called development actors which are the UN agencies, funds and programs, otherwise known as the "UN country team." Unlike primarily military operations (such as those of NATO), many UN peace operations of the last twenty years have often been "multidimensional." They combine a military force, capable in certain cases of carrying out time-bound robust operations at the tactical level, with civilian and police capacities, capable of carrying out peacebuilding and *state building* activities.

The UN force commander of a peacekeeping operation reports to the head of mission,[63] who in turn reports to the Secretary-General through the Under-Secretary-General for Peacekeeping Operations or for Special Political Affairs (for special political missions in the latter case). UN peace operations therefore resemble what the *Stabilization Leaders Forum* (SLF) aspires to in terms of integrated whole-of-government approach combining civil and military tools,[64] the integrated approach the United States and the Western coalition wanted to implement in Afghanistan and after them France in the Sahel by trying to mobilize all the levers and actors involved—security, political or development.[65]

The "integrated" UN approach is not new since it already appeared in Secretary-General Boutros-Ghali 's 1992 *Agenda for Peace*. The latter

[62] Lemaître, Frédéric. "China Wants to Impose Its Vision of Human Rights," Le Monde, November 22, 2019.

[63] with the exception of a few "more military" missions where the force commander is also the head of mission as in UNIFIL.

[64] US Department of State "Stabilization Leadership Forum."

[65] "Emmanuel Macron's Speech For Global French Diplomacy." France UN.

wanted to establish greater unity in the organization's engagement in conflict and post-conflict situations. It is in fact from the United Nations Secretariat that the impetus for better integration of the civilian, military and police components of peace operations came. Quickly, this initiative also sought to strengthen the partnership between peace operations on the one hand and United Nations agencies, funds and programs on the other.[66] Integration first began within the Secretariat headquarters, with the creation in 1992 of the Department of Peacekeeping Operations (DPKO) to support peacekeeping operations which brought together civilian, military, and police forces under the sole leadership of a Special Representative of the Secretary-General and were supported by a dedicated peacekeeping budget.

From the 2000s and following certain recommendations of the *Brahimi report*, new integrated structures were put in place at headquarters and in the field. At the headquarters level, the United Nations adopted an Integrated Mission Planning Process (IMPP) in 2006 to facilitate the planning of multidimensional peacekeeping operations. Permanent Integrated Operational Teams (IOT) were created within the DPKO in order to bring together political, military, police and mission support personnel in a formal structure. The objective was to provide integrated support to missions and an entry point to Member States including troop and police contributing countries for the planning and implementation of so-called integrated peacekeeping operations.[67]

At the level of field operations, a Joint Operations Center (JOC)[68] and a *Joint Mission Analysis* Center (JMAC)[69] were set up that continue to this day, with an equivalent in the special political missions called Joint Analysis and Reporting Units (JARU). This has greatly contributed to better information sharing and joint analysis between the civilian and uniformed components of operations. On the mission support side, Integrated Support Services controlling all logistics resources and Joint Logistics Operations

[66] Arthur Boutellis, "Driving the System Apart? A Study of United Nations Integration and Integrated Strategic Planning", New York: International Peace Institute, August 2013.

[67] UN, "Capstone Doctrine. United Nations peacekeeping operations. Principles and Directions", 2008. P75–78.

[68] Which synthesizes situation reports and operational intelligence from different sources within the mission.

[69] Which provides an integrated analysis of all sources of information to assess threats to the mandate in the medium and long term and to support decision-making by the mission management team.

Centers (JLOC) were also created. In parallel steps to integrate peace operations with UN humanitarian and development agencies, were also taken under the leadership of Secretary-General Kofi Annan, who was convinced that the peacekeeping failures of the 1990s were at least partly attributable to the fact that different components of the UN acted separately, and sometimes at cross purposes.[70]

The first stages of this broader integration included both structural and strategic dimensions, but the most emblematic measure was the creation in 2001 of a "triple-hatted" deputy head of mission position—also Resident and Humanitarian Coordinator—(DSRSG/RC/HC) in integrated peace operations. The DSRSG/RC/HC is responsible for ensuring coordination between mission activities defined by the Security Council mandate and the long-term humanitarian and development initiatives of the United Nations country team whose presence in the country generally predates that of the peace operation.

After these beginnings focused on "structural integration," the focus of the UN integration agenda shifted toward "strategic integration," with a 2008 decision clearly stating that the principle of "integration" must be applied wherever the UN has a country team and a multidimensional peacekeeping operation or a special political mission; and that integration does not only apply to missions that are "structurally integrated" (with a DSRSG/RC/HC) and that it can take different structural forms depending on needs and circumstances.

It is therefore a question of favoring a shared vision of the strategic objectives of the UN in a country between the peace operation and the country team, under the general direction of the head of mission, in particular through common strategic documents.[71] This new approach, more strategic than structural, is summarized in the Policy *on Integrated Assessment and Planning* (IAP) of 2013.[72] The 2019 reform of the UN peace and security architecture also made the new Department of Operational Support (replacing DFS) a service provider for the entire Secretariat (no longer just peace operations) as well as for the Resident Coordinator (RC) system in countries where there are no peace operations.

[70] Kathleen M. Jennings and Anja T. Kaspersen, "Introduction: Integration Revisited," International Peacekeeping 15, No. 4, August 2008, p. 445.

[71] United Nations Secretary-General, "Decision No. 2008/24 – Integration," in Decision of the Secretary-General – June 25 Meeting of the Policy Committee, June 26, 2008, available at www.undg.org/docs/9898/Integration-decision-SG-25-jun-08.pdf.

[72] United Nations, "UN Policy on Integrated Assessment and Planning," March 2013.

However, despite considerable progress and significant innovations in this area in terms of an "integrated" and "multidimensional" approach, the UN has faced significant countercurrents, both internal and external, in recent years. A 2020 review of UN integration concluded that:

> few of those interviewed [for the review] were aware of the Integrated Assessment and Planning [IAP] policy, let alone related tools and techniques ... The two main criticisms of the policy are that associated processes are cumbersome and that it is not relevant for the daily work of staff.[73]

While the production of a joint United Nations *Integrated Strategic Framework* (ISF) by peace operations and the UN country team has in principle been mandatory since 2008, the strategic review of MINUSMA in 2018, for example, noted that there was no such strategic document in Mali at the time. Far from advocating integration, certain United Nations agencies, funds and programs actually sought to distance themselves from the blue helmets who became the target of terrorist attacks in Mali. Some of the UN agencies notably left the MINUSMA super camp in Gao to establish their offices in the town with less perimeter security but explicitly as an attempt to preserve their independence and humanitarian space from a UN force perceived by some as having become party to the conflict under international humanitarian law.[74]

Many Western troop-contributing countries, more accustomed to NATO military structures , "returning" to peace operations in Africa have also questioned in certain cases the "integrated" UN approach. Beyond the fact that some of these countries serving in MINUSMA refused to repaint their planes, helicopters and vehicles white (and preferred to keep the military camouflage), their officers were sometimes resistant to participating in the integrated structures of the mission (the JOC in particular, as well as the aviation component which is traditionally headed by civilian staff with a military officer as deputy). These Western officers often saw these structures and processes as a constraint on the effectiveness of military operations and did not always understand the advantages of the eminently politico-military approach to peace operations.[75]

[73] UN, "2020 Review of UN Integration—Final Report," para. 26.

[74] Arthur Boutellis, "Renewal of the MINUSMA mandate for 2021 in a period of uncertainty", EPON/NUPI, May 2021.

[75] Arthur Boutellis and Michael Beary, "Sharing the Burden: Lessons from the European Return to Multidimensional Peacekeeping," International Peace Institute, January 2020.

The UN has had great difficulty demonstrating the added value of this "integrated" approach that is too often bureaucratized and without real incentives or accountability mechanisms for the leadership. The time spent in coordination meetings and joint strategic planning processes is often considerable. Peace operations lack concrete examples of successful integrated actions, where for example the force could serve as a lever for political actions, or even simply allow the civilian components of an operation to carry out their work. One of the positive examples in MINUSMA has been the deployment of temporary operational bases (TOBs) in particularly dangerous areas of central Mali for months, which allowed civilian components of the mission (such as the civil affairs officers supporting local peace efforts) to carry out activities there.[76]

UN peace operations therefore face some of the same challenges and dilemmas faced by other non-UN stabilization operations. They often favor the strengthening of host state authority (*statebuilding*) over the strengthening of legitimate institutions that can ensure the security of citizens (governance). They often favor the use of military tools to the detriment of more political approaches. They invest in civil-military coordination structures and processes but without this necessarily leading to truly more integrated and effective responses.

In a context of increased rivalries for power and influence, powerful member states have seized on these dilemmas in an attempt to shape UN peace operations according to their respective interests and worldviews. On the one side the P3, certain host countries, and certain troop-contributing countries push for more "robust" and offensive peacekeeping operations, which can use force against certain armed groups or at least support non-parallel forces, and operations that are increasingly ambitious in terms of their mandated activities- from PoC to peacebuilding and stabilization.

On the other hand, China and Russia , as well as many G77 members who are often long-time "traditional" troop contributing countries from Asia and Latin America, instead defend a return to the basic principles of peacekeeping and more modest approaches focused on the respect of the sovereignty and support to the host state, arguing that peacekeeping has become overburdened with peacebuilding tasks and too "intrusive" in the internal affairs and governance of the host state.

[76] Arthur Boutellis, "Renewal of the MINUSMA mandate for 2021 in a period of uncertainty", EPON/NUPI, May 2021.

While "integrated" and "multidimensional" peace operations with an average lifespan of a dozen years seem to have all the tools to implement effective stabilization strategies, they too—like the NATO operations in Afghanistan and French-European operations in the Sahel—have largely failed, but for different reasons. Several factors can explain this failure: a military tool that is too weak and a "use of force" that is poorly controlled[77] and too rarely supports political strategies; development programs that are too inconsequential;[78] and the lack of support from member states and particularly from an increasingly divided Security Council behind political strategies that should guide these peace operations (and even more so when the relationship between the peacekeeping operation and the host state becomes complicated).

Small Political Missions as an Alternative to Large Stabilization Missions?

A major trend of the past two decades, which happen in parallel to the "stabilization turn," has been the multiplication and increasing operationalization of "special political missions" (SPM). These SPMs or political missions have considerably increased in number, size and complexity of the mandates they are asked by the Security Council to carry out. SPMs outnumber the number of peacekeeping operations since 2010.[79]

UN good offices and prevention missions have existed since the very beginning of peace operations. The first dates from 1948 with Swedish Count Folke Bernadotte, the first official UN mediator for Palestine who had helped negotiate the truce before being assassinated in Jerusalem by Jewish extremists on September 17, 1948. Since then, successive Secretaries-General have deployed mediators and other special envoys and personal envoys, either on their own initiative or at the request of the General Assembly or the Security Council.

[77] Due to the very nature of multinational peacekeeping forces, but also to *caveats* imposed by certain troop contributing countries.

[78] In comparison to official development assistance (ODA) provided by States and funding from the World Bank, and the little leverage that the UN has over these despite partnership agreements.

[79] Govinda Clayton, Han Dorussen, and Tobias Böhmelt, "United Nations Peace Initiatives 1946–2015: Introducing a New Dataset," *International Interactions*, June 19, 2020, 1–20, https://doi.org/10.1080/03050629.2020.1772254.

It was not until the 1990s that these types of missions were collectively renamed "special political missions." They were now supported by the new Department of Political Affairs (DPA) created in 1992 in New York and financed by the regular budget of the UN Secretariat rather than the peacekeeping budget.[80] The Secretary-General's first report on the issue in 2015 described them as follows:

> While special political missions vary widely in their functional roles and characteristics, they can be broadly defined as United Nations civilian missions deployed for a limited duration to assist Member States in their good offices, conflict prevention, recovery peace and peacebuilding. They have been grouped into three main categories, or groups, namely special envoys; sanctions monitoring panels and expert groups; and field missions.[81]

But it is under Korean Secretary-General Ban Ki-moon starting in 2007 that political missions and their role evolved most considerably. Ban was initially not particularly interested in peacekeeping operations but wanted to strengthen the capacity of the Department of Political Affairs (DPA) to play its preventive diplomacy role.

The Department of Political Affairs, marginalized under Kofi Annan who had ascended to the post of Secretary-General in 1997 after having headed the Department of Peacekeeping Operations (DPKO) since 1993, then represented only about 200 civil servants. The DPA supported good offices and mediation missions carried out by the Secretary-General and his representatives and envoys, with the support of country analysts. In 2006, a Mediation Support Unit (MSU) was established within DPA and worked closely with its regional divisions to plan and support mediation efforts on the ground.

Its functions include providing advice and support in the context of peace processes, strengthening the mediation capacities of regional and sub-regional organizations, and acting as a repository of knowledge, policies and guidance on mediation. Since 2008, the DPA has also been equipped of a *Standby Team of Senior Mediation Advisers*, a group of experts[82] who can be mobilized "upon request" and can be deployed to

[80] Unlike peacekeeping operations financed by a dedicated budget.

[81] United Nations, United Nations Political Missions: Report of the Secretary-General, July 2015.

[82] With specialist knowledge on issues such as power sharing, natural resources and conflict, the constitution-making process, ceasefires and other security arrangements, and related gender issues to conflicts.

help the Secretary-General's special representatives and special envoys in their functions as mediators in the field.[83]

The majority of the eleven political missions that already existed in 2007 are relatively small, but some, like the United Nations Assistance Mission in Iraq (UNAMI) created in 2003, are larger and made up of hundreds of civilians. The United Nations Assistance Mission in Afghanistan (UNAMA) created the previous year, although similar in terms of format—without Blue Helmets—initially reported to the Department of Peacekeeping Operations (DPKO) before being then managed by the DPA from 2015 onwards. These two UNAMI and UNAMA missions alone represented 3000 employees, or three quarters of DPA's employees and more than half of its budget in 2010.[84] In 2023, UNAMI, UNAMA and the United Nations Assistance Mission in Somalia (UNSOM) were the only three political missions with budgets ranging between $100 and $140 million.[85]

Other political missions consisting mainly of a special envoy and civilian staff were created in 2007 in Lebanon (UNSCOL) and Nepal (UNMIN). Three regional offices were also created, first in West Africa (UNOWA) in the early 2000s which became UNOWAS in 2016 to include the Sahel, in Central Asia in 2007 (UNRCCA), then in 2011 in Central Africa (UNOCA). Finally, medium-sized peacebuilding missions were also tested in Tajikistan, Burundi (BINUB), Sierra Leone (UNIOSIL), and Guinea-Bissau (UNOGBIS) in the 2000s and later in Liberia (UNOMIL) and in the Central African Republic (BINUCA) as well as MINUJUSTH and BINUH in Haiti, which served as successor political missions to the peace-keeping operation MINUSTAH.

The American diplomat Jeffrey Feltman, Under-Secretary-General for Political Affairs of the UN at the time, succeeded in 2012 another American diplomat Lynn Pascoe as the head of the of the Department. The latter, who initiated this reform of the DPA, described political missions whose number had more than doubled[86] the end of Secretary-General Ban Ki-moon's second term in 2015 as follows:

[83] https://dppa.un.org/en/prevention-and-mediation

[84] Richard Gowan (Editor), Review of Politics. Missions 2010. A project of the Center on International Cooperation. New York: NYU Center.

[85] A/75/6 (Sect. 3)/Add.1.

[86] With 23 political missions compared to 16 peacekeeping operations, if we exclude sanctions committees, monitoring groups and special thematic political missions. HIPPO, "Report of the Independent High-Level Panel on United Nations Peace Operations." A/70/95–S/2015/446*". p19.

> The Department [of Political Affairs] manages more than 40 special political missions. These include special envoys tasked with overseeing complex mediation processes, such as in Cyprus, Syria, Somalia and Yemen regional offices in West Africa, Central Africa and Central Asia, which participate in preventive diplomacy on a daily basis alongside regional and sub-regional organizations; field missions with complex and multidimensional mandates to support difficult political transitions in countries like Afghanistan, Guinea-Bissau, Iraq and Somalia; and expert groups monitoring Security Council sanctions regimes. DPA is today more flexible and agile than ever in deploying its know-how, expertise and qualitative advantages in prevention, mediation and peacebuilding in some of the most difficult and intractable contexts of the world.[87]

There is no doubt that this increased presence on the ground revitalized the DPA, and even if most of these missions remained very modest in size, they have the advantage of being inexpensive compared to large peacekeeping operations. The regional offices in particular have proven useful in many ways, for instance in the management of the constitutional crisis and support to ECOWAS mediation in Guinea in 2009, and in the management of the crisis in Kyrgyzstan in 2010. In both cases, the UN played a crucial role in coordinating the positions of the major powers and regional powers involved.[88] The creation of the United Nations Office to the African Union (UNOAU) also responded to the need for increased coordination with the continental organization, whose operation in Somalia (AMISOM) the UN has also supported logistically and financially since 2007. The publication in 2012 of a *start-up guide* for new special political missions by DPA was a strong sign of the institutionalization of this type of missions.[89]

During Ban Ki-moon's first term, the work of two political missions was particularly praised. The one in Nepal (UNMIN) for its support of the democratic transition post-civil war and monarchy. And the one in Iraq (UNAMI) which despite its difficult beginnings with the suicide truck bomb attack against the United Nations headquarters in Baghdad on

[87] United Nations, Department of Political Affairs (DPA), Annual Report Multi-Year Appeal 2015.

[88] France, United Kingdom and ECOWAS in the first case; and Russia, OSCE and European Union in the second.

[89] https://peacemaker.un.org/sites/peacemaker.un.org/files/SPMStartupGuide_UNDPA2012.pdf

August 19, 2003,[90] was then able to play a political role and in particular to ease tensions between the Kurdish and Arab communities, and establish itself as a partner of the new US Obama administration from 2009.

The "peacebuilding missions" in Burundi, Sierra Leone and Liberia, which had a real mandate to coordinate the United Nations system in these countries, and benefited from programmatic peacebuilding funds (PBF), are generally considered successes, although the model has not been replicated since.[91] Other political missions, however, have tarnished the reputation of the UN. In Guinea-Bissau, for example, the organization was powerless when in 2009 the country's president was assassinated, the prime minister was arrested, and soldiers forced their way into UN premises.[92]

From the perspective of Secretary-General Ban Ki-moon, special political missions represented an attractive alternative to peacekeeping operations. In addition to being relatively costly and having difficulties demonstrating their effectiveness, large peacekeeping operations were also the subject of several particularly troublesome scandals during those years—notably repeated sexual abuse scandals in the Central African Republic and the introduction of cholera epidemic by Nepalese peacekeepers into Haiti in 2010, responsible for thousands of civilian deaths.[93]

However, as Richard Gowan notes, DPA reform faces major challenges linked to the lack of interest from the great powers in the majority of these "small missions." Member States in particular refuse to give them adequate and dedicated funding (like for peacekeeping operations), which means that the DPA remains extremely dependent on voluntary or "extra-budgetary" contributions from donors to supplement the regular budget and run and support these field missions:

> Some DPA staff believed that the key to the department's revitalization lay in improving political, technical, and administrative support to these field

[90] Which cost the lives of 22 people including the UN special representative in Iraq, Sergio Vieira de Mello.

[91] The challenge of sustaining peace: Report of the advisory group of experts for the 2015 review of the United Nations peacebuilding architecture, June 29, 2015.

[92] United Nations, "Dismayed by the assassination of a presidential election candidate in Guinea-Bissau, Ban Ki-moon reiterates that impunity must be ended and the rule of law restored in the country", SG/SM/12293, June 5, 2009.

[93] Arthur Boutellis, "Cholera, Haiti and MINUSTAH: What Implications for Peacekeeping?", New York: International Peace Institute, January 11, 2011.

missions—and potentially promoting them as an alternative to DPKO's larger and more expensive peacekeeping operations. But there were considerable obstacles to this project. While some political missions, such as that in Iraq, had a significant profile, many others were weakly staffed and of peripheral interest to the Security Council. DPA officials complained that political missions should be funded from a special budget, as were peacekeeping operations, rather than from the UN regular budget. Given these obstacles, a significant faction in the department opposed concentrating on field missions. Instead, they argued, the priority should be to build up DPA's analytical capacities in New.[94]

Security issues constitute another major challenge for DPA. Indeed, political missions do not benefit from the presence of peacekeepers to stabilize and secure the environment in which they operate. Thus, some of these missions operating in particularly dangerous environments (Afghanistan, Iraq, Libya, Somalia, Syria, and Yemen) have seen their security expenses increase, sometimes representing nearly 40 percent of the mission budget.[95] In Haiti and Afghanistan, UN political missions have employed private armed guards[96] but these practices represented a reputational risks for the UN. In 2013, the United Nations General Assembly formally opposed hiring such private military and security companies (PMSCs), considering they should remain a solution of last resort and only in the absence of alternative, instead e encouraging missions to turn to the security forces of host countries to ensure their protection, to Member States or to the UN Department of Safety and Security (UNDSS). The UNDSS was created in January 2005 following the 2003 attack on the UN headquarters in Baghdad.

Ultimately, the UN developed a model of "guard units" provided by the same troop-contributing countries as for blue helmets and using a similar reimbursement system. The first unit of this type was set up in 2004 in Iraq (with Fijian soldiers). In 2013, three new "guard units" were authorized to protect political missions in the Central African Republic (560 Moroccan soldiers absorbed by the subsequently deployed

[94] Gowan, Richard. "Less Bound to the Desk: Ban Ki-moon, the UN, and Preventive Diplomacy." *Global Governance* 18, no. 4 (2012): 387–404. Accessed November 5, 2020.

[95] International Peace Institute, "UN Peace Operations in Violent and Asymmetric Threat Environments," March 2016.

[96] Colum Lynch, "UN embraces private military contractors," Foreign Policy, January 17, 2010.

peacekeeping mission), Somalia (410 Ugandan soldiers) and Libya (150 to 250 Nepalese soldiers). These units, which are minimal compared to the approximately 100,000 peacekeepers deployed worldwide then, initially wore blue helmets but have since adopted distinct uniforms and insignia.[97] The fact that these units only protect UN premises but do not escort UN staff outside (the responsibility of UNDSS) is however a major limitation.

The United Nations mission deployed in Libya following the fall of Colonel Qaddafi was the real test for the "light footprint" model promoted among others by Ian Martin as he prepared to take the helm of the new United Nations Support Mission in Libya (UNSMIL). His previous experience leading the mission in Nepal (UNMIN) convinced him that "missions should not conform to any model but should be designed according to the particularities of their context, with space to innovate."[98] In the Libyan case, the initial UN plan from August 2011 called for the deployment of 200 unarmed military observers, with a protection force ("such as a multinational force") and 190 UN police officers to help stabilize the country. However, the UN faced strong opposition from the Libyan authorities to the deployment of international uniformed personnel in the final design of the mission deployed to Tripoli.[99]

The Libyan experience raises a double problem. That of security and who can provide these security guarantees for political work in the event that the UN does not deploy its own blue helmets. But also the problem of the limited levers that a political mission has compared to more imposing peacekeeping operation which also generally benefit from stronger support and scrutiny by the Security Council. UNSMIL had to repatriate its employees based in the east of the country in 2012 before the Libyan capital Tripoli for neighboring Tunisia in 2014 in view of the deterioration of the security situation, thus limiting the capacity of the mission to interact with national interlocutors and implement its mandate. There are risks in a return to "minimalist" approaches to peace operations after years of bureaucratic expansion. One of the risks is that operations become simple

[97] "Monthly Forecast: In Hindsight: UN Guard Units," Security Council Report, May 1, 2014.

[98] Ian Martin, "All Peace Operations Are Political: a Case for designing Missions and the next UN reform" in Richard Gowan (Editor), Review of Political. Missions 2010. A project of the Center on International Cooperation. New York: NYU Center.

[99] Arthur Boutellis, "The Test for a UN "Light Footprint" in Libya," IPI Global Observatory, May 21, 2012.

"bureaucrats in suits who read newspapers"[100] providing reports to the Security Council, that these operations are no longer actors but only spectators of developments on the ground. The absence of field offices and specialized civilian employees[101] also necessarily limits the Mission's capacity for analysis and engagement with the parties to the conflict. But it also limits the political and military levers of influence available to these operations, provided the Secretary-General and his special representatives as well as the members of the Security Council are prepared to use these levers.

The United Nations supervision mission in Syria (UNSMIS) deployed in April 2012 without a real ceasefire to be observed and without real consent from the Syrian government itself a party to the conflict, clearly demonstrates the limits of the levers of such operations in contexts where there is no peace to keep.[102] UNSMIS operated for four months, miraculously without casualties among its 300 observers. Given the geopolitical context, a larger, armed and robust peacekeeping force would certainly have been much more exposed (including to terrorist attacks) without necessarily having more leverage. UNSMIS withdrew a few months later in August of the same year following the escalation of the conflict.[103]

In the same way that many of the large "multidimensional" stabilization operations have found themselves having to manage conflicts rather than supporting the implementation of peace agreements, many of the political missions—from Libya to Syria and Yemen—found themselves largely powerless. Rivalries involving great powers and regional powers have often fueled conflicts rather than supported UN peace efforts. Few conflicts today escape power rivalries.

The resignation of the UN special envoy to Libya and head of UNSMIL, Ghassan Salamé, in early 2020 was officially attributed to "health reasons." In reality, the Special Envoy deplored the fact that his mission was

[100] Interview with a UN employee in Haiti denouncing the organizational culture of DPPA following the transition from a peacekeeping operation to a political mission. December 2021. This quote echoes another from *The Fall*, by Albert Camus: "One sentence will be enough for modern man: he fornicated and read newspapers. After this strong definition, the subject will be, if I dare say, exhausted."

[101] In areas such as political affairs, mediation, civil affairs, human rights, protection of civilians, etc.

[102] Arthur Boutellis, "Syria: A UN Observer Mission Like No Other", IPI Global Observatory, 2022.

[103] Richard Gowan, "No Peace To Keep: A Peacekeeping's Year Of Living Dangerously", Worldpoliticsreview.Com, 2013:

torpedoed not only inside Libya, but most importantly from the outside including by UN Security Council members,[104] several of whom ignoring the UN arms embargo imposed on Libya since 2011.[105]

While originally conceived as primarily political, the larger special political mission by virtue of its extended field presence has tended to also expand its remit. UNAMI was closely involved in efforts to limit the suffering linked to the siege of Mosul under Islamic State control, while UNAMA faced the recapture of the country by the Taliban until the fall of Kabul. Independent strategic reviews commissioned by Secretary-General Guterres and transmitted to the Security Council—on UNAMI, UNAMA and UNSMIL—all conclude on the need to reorient these missions to focus on political priorities—that is, peace negotiations—rather than on secondary tasks, including humanitarian ones, which according to these reviews should be entrusted to the UN agencies on the ground.[106]

The UN special envoys for Syria and Yemen also found themselves managing humanitarian challenges linked to the conflict and sometimes de-escalation efforts limited to the local level, without really being able to generate real peace processes. In Sudan, UNITAMS struggled to influence the political transition process after the military junta takeover following the October 25, 2021 coup. The junta threatened to expel UN envoy Volker Perthes, following a critical Secretary-general report to the Security Council at the end of March 2022 and blocked him from briefing the Security Council the following year, prompting the US ambassador to the UN to state: "No country should be able to bully a briefer into silence, let alone the United Nations."[107] UNITAMS was also criticized by the Sudanese pro-democracy movement for having legitimized the coup.[108] Following the resumption of fighting between rival militaries in Sudan,

[104] Carrie Nooten and Frédéric Bobin, "Ghassan Salamé, the UN mediator for Libya, resigns", Le Monde, March 2, 2020.

[105] Le Monde, "Arms embargo violations are increasing in Libya, according to the UN", September 3, 2020.

[106] See United Nations, Security Council, "Report of the Secretary-General on the United Nations Support Mission in Libya," S/2017/726, August 22, 2017, paras. 81–83. And United Nations, General Assembly, Security Council, "Special report on the strategic review of the United Nations Assistance Mission in Afghanistan," A/72/312–S/2017/696, August 10, 2017, p. 5.

[107] Margaret Besheer, "UN Urges Negotiated Solution for Sudan Conflict," VOA, August 9, 2023.

[108] Sudan in the News, "Sudan diplomatic briefing: Challenges facing UNITAMS after the military coup in Sudan", May 20, 2022.

which forced mission personnel to evacuate, the Security Council decided to terminate the mandate of the mission in resolution 2715 of December 2023.

Although limited by their financial and human resources (under the UN regular budget), political missions, some of which resemble "traditional" peacekeeping ceasefire observation missions while being more "multidimensional" with strong civilian staffing,[109] offer today a strong counterpoint to more costly stabilization operations. In a context of exacerbated rivalries between great powers and certain regional powers, the political space of the Secretary-General and his political missions is often extremely constrained. But it is essential for the United Nations Secretariat to try to preserve this space, however limited it may be, if it wants to remain relevant and recognized as such by Member States. This is all the more important as the few major peacekeeping operations are gradually closing—MINUSMA in 2023, MONUSCO likely in 2024—with no large peacekeeping operation authorized since 2014 and no new ones in sight.

Liberal "Status Quo" Versus "Revisionist" Powers

In an interview with the newspaper *Le Monde* at the end of June 2021, Secretary-General, Antonio Guterres, recently reappointed to his post for a second term until 2027, suggested that the Security Council paralysis gives "greater room for maneuver to middle powers."[110] In this he echoes Secretary-General Dag Hammarskjöld, who in the midst of the Cold War, declared that "it is not the Soviet Union or any other great power that needs the United Nations for its protection: it is all the others."[111] Dag Hammarskjöld was, in fact, able to count on the support of certain middle powers, while the two blocs neutralized each other.

[109] Alexandra Novosseloff, "A comparative study of older one-dimensional peace operations: Is the Future of UN Peacekeeping its Past?" EPON/FES, April 2022, See also Zachary Myers & Walter Dorn "UN Peacekeeping Missions in the Middle East: A Twenty-First Century Review", International Peacekeeping, 2022, 29:3, 413–435, DOI: https://doi.org/10.1080/13533312.2022.2053287.

[110] Carrie Nooten, "Antonio Guterres: "We lack instruments to combat new conflicts" For the UN Secretary-General, the paralysis of the Security Council gives "greater room for maneuver for middle powers"," Le Monde, June 30, 2021.

[111] Dag Hammarskjöld, Statement before the General Assembly, October 3, 1960 (A/PV. 883).

The special status of "middle power" at the UN is a status claimed by Canada from the beginning of the organization at the San Francisco conference, promoting a multilateralism reflecting the ideals of an "international liberal order."[112] A few years later, in 1956, Canada sent a thousand soldiers to the first peacekeeping force in the Sinai in the context of the Suez Crisis. RG (Gerry) Riddell, a Canadian diplomat who worked with Canadian Foreign Minister Lester B. Pearson, already defined "middle powers" in 1947 as countries that:

> because of their size, their material resources, their will and their ability to assume their responsibilities, their influence and their stability, are close to the great powers.[113]

Great powers often have an ambivalent relationship to the UN. They are proud of their P5 status and use their power to obtain the support of other member states, but do not hesitate to use or even abuse the UN Secretariat, or to bend the rules to assert their interests and even in rare cases to bypass the Security Council to intervene militarily. Conversely, middle powers tend to respect the rules and want to preserve a certain *status quo* in the multilateral system which they see as regulating the *hubris* of the great powers and the anarchy of the international system. This includes support for UN peacekeeping, which they consider a "global public good." Peacekeeping is also one of the classic contributions of middle powers to multilateralism, including for Western middle powers when they run for an elected Security Council seat.

Professor John Ikenberry speaks in this regard of states that are strong multilateral "institution builders," which agree to limit the use of their power, in order to obtain from middle powers and smaller states that they participate and comply with the institutions set up in the interest of the strongest.[114] But that balance is being challenged today as member states are divided at the UN and major powers including regional powers appear less and less willing to voluntarily limit their power. In this context, certain

[112] John Ravenhill, "Cycles of Middle Power Activism: Constraint and Choice in Australian and Canadian Foreign Policies," *Australian Journal of International Affairs* 52, no · 3 (November 1998): 309–27, https://doi.org/10.1080/10357719808445259.

[113] RG Riddell, cited in RA MacKay, "The Canadian Doctrine of Middle Powers," in HL Dyck and HP Krosby (eds) *Empire and Nations*, Toronto, University of Toronto Press, 1969, p. 138.

[114] Ikenberry, *After Victory*.

UN peace operations such as MINUSMA have also ended up becoming a field (among others) of confrontation in a context of "crisis of multilateralism" whereby the "liberal international order" and certain spheres of influence are being challenged by certain influential member states.

Many middle powers have been concerned in recent years about the disengagement of the United States under President Trump, the blockages in the Security Council, and the almost systematic vetoes by Russia (and sometimes China) on Syria.[115] They are also more widely concerned that major and certain regional powers increasingly appear to defy international rules and norms, with violations of international humanitarian law in Syria and Afghanistan (including attacks on schools and hospitals) notably, Russia's annexation of Crimea in 2014, attempts to interfere in the US presidential elections, the militarization of the China Sea, and the invasion of Ukraine in February 2022.

The fact that great powers have largely disinterested themselves from maintaining peace in countries not directly relevant to their national interest has created for other member states, some of which have promoted an international liberal order. This partly explains the rapid expansion of Protection of Civilians and human rights standards in peace operations described above. Mélanie Albaret explains that:

> the UN offers a few States, those which consider themselves strong enough to participate in defining the rules of the international game but not strong enough to act alone, the possibility of engaging in an original way. It reduces the costs of collective action and provides these States with at least symbolic rewards by giving them an international role. The UN raises the international status of states that engage in it by establishing middle power strategies. Consequently, this relationship allows the reversal of the original relationship between the UN and the power: the first can bring about the second. These middle-power multilateral strategies are diverse, sometimes complementary and evolving. They can take the form of building and leading multilateral coalitions, niche diplomacy (Andrew F. Cooper), developing an image of a good member of the international community, etc. In these multilateral strategies of middle power, the UN occupies a privileged place because it is broader than regional forms of multilateralism (European Union, MERCOSUR, ASEAN, etc.) and than select clubs (G8, G20, BRICS). It remains the only universal organization with general compe-

[115] Since 2013, France has supported with the United Kingdom the regulation of the use of the veto for situations of mass atrocity.

tence. As a result, the various initiatives at the UN easily acquire a global dimension, can benefit from significant media coverage and therefore confer on the States which carry them visibility and importance which helps their capacity to do, to get done or to prevent.[116]

For many middle powers, peace operations have thus become a means of demonstrating their commitment to support multilateralism and collective security, either through troop contributions, even modest ones, through extrabudgetary financial contributions, or through contributions to various policy discussions in New York. But it is also a way for middle powers to raise their status, which was one of the reasons behind the re-engagement of certain Western troop contributors in peacekeeping in MINUSMA.

For instance, a country like The Netherlands both contributed peacekeepers to MINUSMA in 2014 in the context of its candidacy for non-permanent seat on the Security Council but also continued supporting policy discussions on peacekeeping in New York. As Niels van Willigen explains, "the decision to participate in MINUSMA was mainly explained by national interests and internal factors."[117] The Dutch commitment as one of the most active "champion" of the "Action for Peacekeeping" (A4P) Declaration of September 2018 (and again in 2021 for the launch by the Secretariat of United Nations of A4P+[118]) can also be explained by the desire to assert its status and "prolong" its influence and visibility beyond its short stint on the Security Council in 2018.[119]

As Richard Gowan succinctly summarizes, the A4P Declaration reflects support for a set of "pre-existing ideas" rather than "new ideas about how states should cooperate."[120] For The Netherlands as for many other middle powers it is therefore a question of defending a certain *status quo* on peace operations, in a context where Member States seem to be

[116] Mélanie Albaret, "The UN, between power and multilateralism", CERISCOPE Puissance, 2013.

[117] Niels van Willigen, "A Dutch Return to UN Peacekeeping?", *International Peacekeeping* 23, no. 5 (October 19, 2016): 702–20, https://doi.org/10.1080/13533312.2016.1235095.

[118] United Nations, Peacekeeping, "Three years of Action for Peacekeeping (A4P): Achievements, Challenges and the Way Ahead," March 29, 2021.

[119] Since the Pas Pays and Italy have decided to share their non-permanent headquarters in 2017–2018.

[120] Richard Gowan, "The End of a Peacekeeping Era," Global Peace Operations Review, April 4, 2018.

disinterested in UN peacekeeping, which is also under attack from the US administration of President Trump. The Netherlands also promoted certain of its own priorities such as intelligence in peace operations and the protection of civilians.

Middle powers neither have permanent seats on the Security Council nor "reserved" Under Secretary-General positions in the UN Secretariat. But many of them share some of the same ideas—which makes them *like-minded*—about the importance of multilateralism and try to build alliances (*bridge-building*) to gain legitimacy and influence at the UN despite their limited means. A preferred mode of action consists in providing extrabudgetary financial contributions to specific projects at the UN Secretariat and in field operations, but also financing the "Third UN " of *think tanks* and NGOs, to try to also influence indirectly the system.

The all-out German extrabudgetary financial contributions in recent years—both upstream in support of their campaign for non-permanent seat on the Council in 2019–2020 and downstream to assert their legacy—undoubtedly helped the country secure leadership positions in the UN. In January 2021, Volker Perthes was appointed head of the new political mission in Sudan UNITAMS and in January 2022, Daniela Kroslak was appointed deputy head of MINUSMA; while the German former head of UN Environment was already at the head of UNDP since 2017. Germany also used its seat on the Council to promote the role of women in peacekeeping[121] and to support the efforts of the special envoy for Libya through the Berlin process, the objective of which was to reduce external influences and thus enable an intra-Libyan political process under the auspices of the UN.[122]

The United Kingdom and France, members of the P5, still hold the pen in the Security Council on certain situations reflecting their strategic interests and their historical (and colonial) affinities, but are today widely considered as "powers that have become middle"[123] or "middle powers with global ambition."[124] Because of that, their behavior in the Security Council is evolving. Since 2019, the United Kingdom has symbolically

[121] Hannah Birkenkötter, "Taking stock: A review of Germany's two years on the Security Council", Völkerrechtsblog, January 25, 2021, doi: https://doi.org/10.17176/20210125-191100-0.

[122] Weitere, "Over 100 resolutions later: A look back over two years in the UN Security Council," December 23, 2020.

[123] Charillon, *France in the world*.

[124] Badie, *France a thwarted power*. P10, p33, p94, p244.

offered certain elected members, including Germany, to become co-pen on certain issues. In the context of *Brexit*, the United Kingdom seems to assume the role of "middle power" supportive of multilateralism, deploying quality troops under blue helmets first in South Sudan from 2017–2020 then in Mali in 2021 and communicating abundantly on their actions to protect civilians. But through these actions the United Kingdom also seeks to reaffirm its international legitimacy in connection with its status as a permanent member of the Security Council.

In Asia, Japan and the Republic of Korea (RoK), respectively in third and tenth positions among contributors to the budget for peacekeeping operations, have also adopted "responsible middle powers" and constructive strategies at the UN. Both countries are constrained by their proximity to the United States and worried about the growing power of their Chinese neighbor, including in the multilateral system. They are cautious in their contributions to peace operations which consist mainly of military engineering units but are masters at organizing large international conferences which give them a certain international prestige. RoK for instance hosted the annual UN Peacekeeping Ministerial on December 7, 2021 in Seoul (and launched the "Seoul Initiative on strengthening technological and medical capacities in peacekeeping"), which positioned it well to champion peacekeeping while serving as a non-permanent member of the Security Council in 2024–2025 (and chair of the Security Council Working Group on Peacekeeping Operations, a mechanism established in 2001 to improve cooperation between the Council and troop-contributing countries).[125]

The concept of "middle power" refers to "states that are neither large nor small in terms of international power, capacity and influence, and which demonstrate a propensity to promote cohesion and stability in the global system."[126] However, in recent years and in a context of power rivalries, some of the P5 members that could be considered larger middle powers in economic terms and in terms of military expenditures—in comparison to the two great powers, the United States and China—but which maintain a global ambition,[127] have increasingly weighted onto the trajec-

[125] https://peacekeeping.un.org/fr/2021-seoul-un-peacekeeping-ministerial

[126] Eduard Jordaan, "The Concept of a Middle Power in International Relations: Distinguishing between Emerging and Traditional Middle Powers," *Politikon* 30, no 1 (May 2003): 165–81, https://doi.org/10.1080/0258934032000147282.

[127] Justified in part by their permanent seat on the Security Council in the cases of France and Russia but also their military "adventurism".

tory UN peace operations have taken. The case of MINUSMA is a good illustration of how France has in some ways been "disruptive" of the institutional *status quo* on UN peace operations, while Russia has played more of a "revisionist" and even "spoiler" role, despite efforts by other Council members to limit its feared spoiler role on peace operations following the 2022 Ukraine invasion.

France can be considered "disruptive" to the peace operations *status quo* in that while defending its special P5 status and a rules-based international system, it has attempted to move UN peacekeeping toward greater robustness (and therefore use of force) and to involve the peacekeepers in support of the fight against terrorism in Mali to advance its national interest and support those of its allies. This manifested itself when France supported the Mali government request that the MINUSMA mandate be made more robust in 2016, when it advocated for greater UN support to the G5 Sahel Joint Force (even if the United States and the United Kingdom largely neutralized the French initiative), but also through some of the actions by French officers placed in key positions within the military component of MINUSMA. France of course also benefited from "occupying" the position of Under Secretary-General for Peacekeeping—notably when it came to neutralizing some of the recommendations made by the 2018 independent review. It has also benefited to some degree from the position of Secretary-General Antonio Guterres who has often defended the idea of more comprehensive support from the UN to non-UN peace enforcement forces, including counterterrorist, authorized by the Council under chapter VII.[128]

Russia, at the difference of France, has long disengaged from UN peace operations. The experience of the Balkans and Kosovo has shaped Russian perception that UN peace operations are manipulated by the West, and this has partly motivated its decision to deploy its own troops within the framework of the Collective Security Treaty Organization (CSTO), an intergovernmental military alliance in Eurasia consisting of six post-Soviet states created in 2002, and which protects Russian interests in its neighborhood. Activities of CSTO "peacekeepers" range from guaranteeing a ceasefire in Nagorno-Karabakh to supporting an allied regime in Kazakhstan. Russia has therefore not tried to disrupt or shape the *status quo* around UN peace operations but has instead played more of a "revisionist" and "spoiling" role in recent years. According to political scientist

[128] RFI. 2018. A. Guterres (UN): "A clear mandate and funds for African forces."

Randall Schweller, at the difference of "status quo states [which] are content to preserve the essential features of the existing international order and the general distribution of power," "revisionist" powers like Russia, dissatisfied with the international order, will not hesitate to use military force to change, not to preserve, the international status quo and extend their values.[129]

Russia in recent years has thus actively discredited on the ground the same UN peacekeeping forces whose mandates it authorized in the Security Council, in Africa in particular. *Wagner Group* mercenaries in the Central African Republic then in Mali, accompanied by disinformation campaigns, have greatly contributed to discrediting the former colonial power France in the eyes of host country governments and populations, but also by extension UN peacekeepers and the UN more generally including its Security Council where France holds the pen on the situations in these countries. Obstructions and restrictions on air and land movements in particular became common and peacekeepers have increasingly been accused of passivity in the face of armed groups (when *Wagner Group* mercenaries were active alongside national forces). UN operations have always been easy scapegoats, but this behavior by a member of the P5 will certainly have deleterious impact UN peacekeeping in the longer term.

Russia, like China in Asia, has always preferred to keep the world organization and its peacekeepers at a distance from what it considers to be its ex-Soviet precinct. Russia also benefits from a privileged relationship with China when it comes to votes in the Security Council. But unlike China which asserts its power and seeks to transform and shape the UN according to its own model, while also supporting the multilateral system and notably its peacekeepers, Russia's actions on the ground and in New York focus more on securing votes in the General Assembly, as illustrated in the March 2, 2022 vote on a resolution "demanding that Russia immediately withdraw its forces and abide by international law," on which a certain number of countries hosting UN peace operations and *Wagner Group* mercenaries—including the Central African Republic, Mali and Sudan—as well as China abstained.[130] China allegedly played an important role in making sure Russia wouldn't play spoiler in subsequent Council votes

[129] Randall L. Schweller, "Bandwagoning for Profit: Bringing the Revisionist State Back In," International Security Vol. 19, No. 1 (Summer, 1994), pp. 72–107.

[130] Le Monde, "War in Ukraine: massive vote at the UN for the withdrawal of Russian troops", March 3, 2022.

renewing peace operations—in South Sudan[131] and Afghanistan notably—and making sure Russia and Western P3 powers continue to "compartmentalize" their differences, but it was however not able to prevent the end of MINUSMA—despite China having its own peacekeepers deployed in that mission—(and UNITAMS) in which Russia likely played a considerable role.

Middle powers, with the exception of a few "rogues", mostly serve the agenda of UN peace operations, ostensibly to try to preserve this important tool of multilateralism, but often also to try to shape it so that it reflects their values and can in one way or another serve their interests. These middle powers "use" peace operations and the UN more broadly to obtain non-permanent seats on the Council, gain influence, international legitimacy, prestige and status (or attempt to limit their decline) especially when they do not have the economic and military means to play more coercive roles. The room for maneuver of these middle powers vis-à-vis the great powers is often limited but they can collectively—within the E10 in the Council and through cross-regional like-minded groups in the General Assembly—influence the implementation and overall trajectory of UN peace operations. While certain initiatives, notably from E10 members remain more at the "micro" level, aiming to improve the functioning of its operations,[132] others led by "coalitions" of middle powers do impact the "macro" level, to use the terminology used by Roland Paris.[133]

However, as Roland Paris points out, these initiatives which struggle to go beyond the reaffirmation of principles cannot alone safeguard "the liberal international order" and "no system of international institutions and rules can survive very long without the support of its most powerful members." He adds that "the challenge, moreover, is not simply to preserve key elements of the existing order, but also to reform institutions that are failing, devise new rules for emerging policy areas, and adapt the multilateral system so that it better reflects the shifting realities of global power, rather than the post-1945 world in which its central institutions took shape." And to conclude that "the collective influence of the middle

[131] This is despite Russia and China attempting to remove previously agreed language on issues such as climate change, human rights, and women, peace and security, and abstaining from these. Two countries. See Security Council Report. "UN Mission in South Sudan (UNMISS): Vote on Mandate Renewal Resolution*" *What's In Blue*. March 14, 2022.

[132] Like bringing normative language into a resolution on peace operations.

[133] Roland Paris, "Broadening the Study of Peace Operations," *International Studies Review* 2, no 3 (December 2000): 27–44, https://doi.org/10.1111/1521-9488.00214.

powers should not be underestimated" because if they do not mobilize, no one will do it for them.[134] Indeed, as Guillaume Devin reminds us, "all powers, large, medium and small, share this tendency to exploit multilateral institutions: to use them especially if they can serve national interests."[135] In a context of global disorder, middle powers today face dilemmas and adopt different strategies to promote both their values and their interests. Some choose to defend the institutional *status quo* while others seek to define the peace operations of tomorrow. They act collectively or individually, to try to move the lines on UN peace operations, which are all the more flexible since they are not enshrined in the Charter of the United Nations.

UN peace operations and the multilateral system more broadly are not a simple reflection of inequalities between States, and middle powers can and sometimes do resist the most powerful. They seek to preserve and influence the direction of these operations, either collectively particularly in the Security Council, and power games within the UN bureaucracy; or individually by seeking to constrain certain operations from the outside through a "parallel force," which can take radically different forms, from the Council-authorized French operation in Mali to Russian *Wagner Group* mercenaries. The trajectory of peace operations and their future can therefore only be understood if we take into account the changing geopolitical balance of power and the more or less well-defined power strategies of key member states. In a context where the most important multidimensional UN stabilization operations are losing momentum and more modest political missions in various formats are on the rise, middle powers can no longer be content with contributing few troops and military observers under helmet or blue beret and need to develop more robust, collective, diversified but comprehensive influence strategies both at the level of headquarters in New York and on the ground where UN field operations are deployed.

[134] Roland Paris, "Can Middle Powers Save the Liberal World Order?", Chatham House, June 18, 2019.

[135] Guillaume Devin, "The future of multilateralism: Why is multilateralism resilient and fragile despite everything?", Conference at the Henri La Fontaine Foundation (Brussels), January 22, 2020.

CHAPTER 6

The Return of Power Rivalries and the Future of UN Peace Operations

"The UN is not just a product of do-gooders. It is harshly real. The day will come when men will see the UN and what it means clearly. Everything will be all right — you know when? When people, just people, stop thinking of the United Nations as a weird Picasso abstraction, and see it as a drawing they made themselves."

Dag Hammarskjöld, *The Times [London]*, 27 June 1955

Until relatively recently, UN peace operations had been largely shielded from power rivalries and Security Council divisions, generally between the US—and France and the UK—on one side (the P3) and Russia and/or China on the other side (the P2), over issues such as Syria or Ukraine. Phillipe Cunliffe even suggests that the significant degree of post-Cold War international cooperation around peacekeeping issues challenges conventional realist assumptions about the selfish nature of states.[1] Until 2017 and the arrival of Secretary-General Antonio Guterres at the head of the world organization and simultaneous arrival of President Donald Trump in the White House, members of the Security Council had indeed demonstrated a unique ability to "compartmentalize" their differences, and thus maintain a common front on the deployment of peace operations in Africa

[1] Oksamytna and Karlsrud, *United Nations Peace Operations and International Relations Theory*. p. 39.

A. Boutellis, *The UN Stabilization Mission in Mali (MINUSMA)*, Twenty-first Century Perspectives on War, Peace, and Human Conflict, https://doi.org/10.1007/978-3-031-66466-3_6

notably, where most contemporary operations have been. The year 2018 however seems to have been a turning point for UN peace operations, which have become yet another terrain of power rivalries. The arrival of the Russian *Wagner Group* in Africa around 2018 in Central African Republic (and later in 2021 in Mali) and the invasion of Ukraine in 2022 further exacerbated these rivalries.

Over the past six years, since 2018, the UN started to feel the consequences of these power rivalries within the Security Council and beyond, and peace operations are on the way to becoming an important, albeit still mostly indirect, field of confrontation between major and middle powers in a context of questioning of multilateralism and the "international liberal order" perceived by a growing number of member states as a "Western order." These tensions started contaminating debates on mandates which were largely consensual until now, as we see during the final phase of MINUSMA described in Chap. 4 of this book. The consequences of power rivalries are felt beyond the Security Council, in the General Assembly but also the United Nations Secretariat (or the bureaucracy), through negotiations of Council documents, budgets, but also in the field of principles and norms, as well as in increasingly politicized senior appointments. They also manifest themselves directly in field operations through troop contributions, the growing assertiveness of host countries, and the behavior of various "parallel forces"—including but not only the *Wagner Group*, which also impact UN operations directly on the ground.

Growing Security Council Divisions

The first set of consequences is fault lines among the permanent five (P5) members of the Security Council. Of the P5, the United States, United Kingdom, and France (P3) have traditionally led the drafting of Council resolutions mandating peace operations, while Russia and China (P2) have long shared their disagreement with the interpretation of the core principles guiding peacekeeping. But these differences had however not really yet affected the ability of the fifteen members of the Council to agree on language and unanimously vote peace operations mandates, whether the authorization of a new operation or the renewal of the mandate of an existing one. Over the past few years however, these differences have started seeping into both thematic debates and mission-specific discussions. This was clear in September 2018 when Russia circulated a draft presidential statement which deleted twenty paragraphs from a draft resolution the US shared on peacekeeping performance during its presidency. Also, for the first time in 2018, four peacekeeping missions did not enjoy unanimously supported mandates (when in the past unanimous votes on peace operations were almost taken for granted), and this trend continued and sharpened the following years.

Russia, which is showing renewed interest in Africa and deployed the *Wagner Group* and military trainers to the Central African Republic (CAR)[2] and other African countries, started challenging France on peace operation in its former colonies. The disagreement, based on Russia's fear that the mandate renewal process would not recognize its mediation efforts with the Sudanese as part of the "Khartoum process," delayed the renewal of the mandate of MINUSCA and in March 2019, Russia abstained on the new mandate for the United Nations Mission in South Sudan (UNMISS) this time. The *International Crisis Group* also noted a more systematic Russian opposition to initiatives of the P3[3] which was confirmed by Russian (and Chinese) abstentions in November 2021 on the renewal of the MINUSCA mandate and in June 2022 for the first time on the renewal of MINUSMA's mandate. The United States had notably called on Russia to investigate allegations of abuse committed by "Russian contractors" (understood as *Wagner Group* mercenaries).[4] The following year in June 2023, the representative of Russia asserted that "any proposals for MINUSMA's mandate revisions should be based on the host country's opinion", thereby anticipating on the Malian representative asking for the "withdrawal, without delay" of MINUSMA when no other Council member nor the UN seemed to have been warned of this sudden decision.[5]

The rivalry between the US and China is also increasingly playing out over peacekeeping mandate renewals. In 2014, as China began significantly stepping up its contributions to peacekeeping by sending its first infantry units (when it had previously focused its contributions on medical, engineering, and logistics) to South Sudan and Mali, Beijing had requested that the UNMISS mandate include a reference to the protection of oil infrastructures. The US initially resisted, but eventually accepted the reference, recognizing China's growing role as a power

[2] Nathalia Dukan, "Central African Republic: Ground Zero for Russian Influence in Central Africa", Issue Brief, Atlantic Council, October 2020.

[3] Crisis Group, "Council of Despair? the Fragmentation of UN Diplomacy".

[4] Michelle Nichols. "U.S. Confronts Russia as U.N. Renews C.Africa Peacekeepers." Reuters. Thomson Reuters, November 13, 2021.

[5] Security Council 9350th Meeting (am), Sc/15325, June 16, 2023.

broker in South Sudan, and thereby accepting a more transactional relation.[6] In 2019, the US blocked a regular renewal (instead granting a "technical rollover" for six months only) of the mandate of the UN Assistance Mission in Afghanistan (UNAMA) over a reference to China's *Belt and Road Initiative* (BRI), which China had managed to get into the resolution the previous year. The representative of Germany (the co-penholder on Afghanistan together with Indonesia that year) publicly regretted that "issues that have nothing to do with UNAMA's mandate emerged, making it impossible to pass the resolution that the penholders sought."[7]

During the renewal of the mandate of the United Nations Integrated Office in Haiti (BINUH) of October 2021, China, unpleased with the country's ties with Taiwan and insisting on the primary responsibility of the Haitian government in the management of its challenges, did not did not hesitate to prepare a counter-proposal for a resolution for six-month only, before accepting a nine-month compromise (with a strategic review after six months) by the American and Mexican co-pens.[8] And during the July 2022 mandate discussions amid rising gang violence, China was once again very active, stepping outside of its traditional stances to propose an arms embargo and possible sanctions under Chapter VII. China even proposed sending a multinational police unit to Haiti to fight against gangs.[9] The United States blocked the proposal which directly called into question its management of the Haitian crisis in recent years—all the more that the United States has imposed two successive American candidates to head the peace operation since 2017.[10] The US instead started working bilaterally with Kenya on a project for a non-UN "multinational force" to Haiti—in part at least to block possible China influence—which it

[6] Lynch, Colum. "U.N. Peacekeepers to Protect China's Oil Interests in South Sudan." Foreign Policy, June 16, 2014.

[7] "Security Council Extends Mandate of United Nations Assistance Mission in Afghanistan, Unanimously Adopting Resolution 2460 (2019) | Meetings Coverage and Press Releases." United Nations.

[8] "Haiti: Last Friday's Vote on the UN Integrated Office in Haiti (BINUH) Mandate Renewal." Security Council Report, February 2022.

[9] Security Council Report, "Haiti: UN Integrated Office in Haiti (BINUH) Mandate Renewal", July 15, 2022.

[10] Nations Unies, "Secretary-General Appoints Susan D. Page of United States Special Representative, United Nations Mission for Justice Support in Haiti", 1 November 2018, SG/A/1767*-BIO/5039*-PKO/683*.

introduced to the Council in Resolution 2699 (2023), which passed by a vote of 13 in favor but with 2 abstentions by China and Russian.[11]

China and Russia have also expressed their misgivings regarding the penholder system and warned against "steamrolling divisive draft resolutions."[12] Russia even organized an informal *Arria formula meeting* on pen holding on August 11, 2022. This was an opportunity for the P2 and elected members of the Council to denounce the quasi-monopoly of the P3 on the pen-holding system since the early 2000s.[13] Since the pen-holding system is an informal arrangement, any Council member can in theory be a penholder but if China were to request to hold the pen on more files in the future, deadlocks between the US and China could become a regular occurrence. Penholding has been defined by the Security Council as "the informal arrangement whereby one or more Council members (as 'penholder(s)') initiate and chair the informal drafting process."[14] But penholding remains contentious given both the perception and reality of the influence it gives a Council member over the issue under consideration. In addition to leading negotiations and the drafting of outcomes and therefore having the opportunity to shape language and drive action, penholders also get to speak first during the Council's formal and informal meetings.

More surprisingly, new divisions have also emerged between traditional allies like the US and France, notably during the Trump administration years, over the cost of peacekeeping in Francophone Africa—from Mali, to CAR, and the Democratic Republic of Congo (DRC). In 2018, the US blocked the renewal of MINUSMA, with the Council ultimately reaching a compromise over a six-month review and new benchmarks. During the 2019 MINUSMA renewal, the US pushed for the redeployment of peacekeepers from northern Mali to protect civilians in the center of the country, but within existing resources, when France argued that this risked allowing jihadists back into the north but eventually agreed to the adaptation in the mandate. The US-France tensions eased when the Biden administration took office but left some scars.[15]

[11] Security Council, Resolution 2699 (2023) adopted by the Security Council at its 9430th meeting, on 2 October 2023.

[12] Security Council Report, "2022 Chairs of Subsidiary Bodies and Penholders", February 2022.

[13] Security Council Report, "Arria-formula Meeting on Penholdership", August 10, 2022.

[14] UN Doc. S/2017/507, para. 78.

[15] "Mali: Renewal of MINUSMA's Mandate*." Security Council Report. June 28, 2021.

The ten non-permanent or elected members of the Council (E10) have also increasingly coordinated and asserted their positions since 2017–2018 in a context of worsening P5 tensions and the wider erosion of the Council's credibility, notably over its failure to act in Syria.[16] In 2019, Germany becoming co-pen holder on the peacekeeping mission in Darfur as an E10 member was a modest but symbolic evolution and other similar co-penholderships have happened on peace operations since. However, France reportedly turned down Germany's suggestion to co-pen the resolution on MINUSMA. The E10 indeed continue to face resistance from some P5 members who are more jealous than others of their prerogatives, including the unwritten ones. Russia and the United States, for example, joined forces to block a resolution on improving the mandates of peacekeeping operations tabled by Côte d'Ivoire and the Netherlands in November 2018 as part of the Declaration "Action for Peacekeeping" (A4P).[17]

Using and Abusing Financing Leverage

The financial issue—discussed in the fifth committee of the UN General Assembly responsible for administrative and budgetary matters—has been another important leverage used by various member states at the UN. While financial crises at the UN tend to be cyclical, the latest started in 2017 when the newly elected Trump administration became more assertive in its efforts to cut UN peace operations costs. In the first quarter of 2017, the new permanent representative of the United States to the UN, Nikki Haley, threatened to boycott the renewal of the mandate of MONUSCO unless there were deep reductions. She was finally satisfied with a modest reduction in the number of peacekeepers. Interestingly, the budget is not only used by the biggest financial contributors to the Organization's regular and peacekeeping budgets, namely the United States (which pays approximately twenty five percent) and China (which pays approximately eighteen percent). Russia also attempted to use financial pressure in the fifth committee of the General Assembly in relation to the June 2023 MINUSMA withdrawal resolution (initially asking that the

[16] See Arthur Boutellis, "Lessons from E10 Engagement on the Security Council," International Peace Institute, November 2022.

[17] "Is Christmas Really over? Improving the Mandating of Peace Operations: Research Report" Security Council Report.

organization only fund a three-month instead of a six-month withdrawal), similarly to what the US administration had done when it launched "second battles" over peacekeeping budgets when it couldn't get the cuts it had hoped for in mandate renewal discussions in the Security Council.[18]

Despite significant reductions in the UN peacekeeping assessed budget in the past few years due to the closing of multiple peace operations, cash-flow problems related to the late payment and withholding of assessed contributions by the US placed a new strain on peacekeeping. This undermined the delivery of mandates and reform efforts, and the timely reimbursement of troop- and police-contributing countries (TCCs/PCCs) and has strained Security Council-led discussions on performance. On January 26, 2024, the Secretary-General sent a letter to the UN staff on the UN financial situation, explaining that the main cause of the liquidity crisis is that not all Member States pay their assessments in full and that in 2023, "only 82.3 per cent of the year's assessment" was collected and only 142 Member States out of 193 paid their dues in full, the lowest in the last five years; "as a result, year-end arrears climbed to $859 million, up from $330 million in 2022." As a consequence, the UN Secretariat has had to impose spending and hiring restrictions (personnel costs accounting for more than 70 percent of the regular budget).

Interestingly, evolving geopolitics and the renewed interest of Western powers toward "Global South" priorities, has led the US Biden administration to champion in 2023 a resolution on the financing African Union(AU)-Led Peace Support Operations authorized under Chapter VII,[19] when in December 2018 it was the (previous) US administration which had blocked a resolution proposed by the three African elected members of the Council (A3), calling on the UN to fund AU peace operations due to its possible financial implications. Indeed, the P3, China and Russia have been competing for influence on the African continent (representing 54 member states and as many votes in the General Assembly), where many countries have refused to take side on the war in Ukraine. On December 21, 2023 the Council finally passed Resolution 2719 (2023) on the financing of AU-led peace support operations (AUPSOs), authored by the Council's three African members (Gabon, Ghana, and Mozambique,

[18] Arthur Boutellis, "The Threat of US Cuts: Helping Peacekeeping Help Itself?" The Global Observatory, March 30, 2017.

[19] Dawit Yirga, Paul Romita and Karin Landgren "The United Nations in Hindsight: Financing African Union-Led Peace Support Operations," Just Security, April 2, 2023.

also known as the A3).The need to appeal to "Global South" countries while also countering the increasing tendency by some African countries to rely on *Wagner Group* mercenaries, clearly helped the United States in particular overcome some of its (and US Congress) previous concerns over accountability and oversight[20] even if the US Biden administration has since lost much of its diplomatic standing at the UN due to its positions on Israel and the war in Gaza.[21]

In recent years, the voluntary contribution of "extra-budgetary" funding to compensate for the shortcomings in obligatory contributions to the regular and peacekeeping budgets of the UN have also increasingly become a way for member states to try to influence the system. Special political missions and the Department of Political and Peacebuilding Affairs (DPPA) are particularly dependent on such voluntary contributions, much of which comes from Western middle powers.[22] These same countries also contribute significantly to the financing of programs and projects implemented by peace operations on the ground, either through the peacebuilding fund[23] or more directly through contributions to mission *trust funds*. For instance, over the years the trust fund set up by MINUSMA had received more than 30 million from Germany, 12 million from Denmark, 9 million from the Netherlands and Canada, etc. Extrabudgetary (XB) voluntary contributions have become essential to the functioning of the UN system and sometimes encourage innovation but—along with Junior Professional Officers (JPO) and *gratis personnel* sponsored certain member states[24]—are often perceived as indirect ways for the wealthiest member states to advance their agendas and some policies outside the formal framework of the General Assembly committees (the Special Committee on Peacekeeping Operations "C34" in particular). A striking example has been how the C-34 explicitly limited the use of 'intelligence" in peacekeeping contexts to the sole purpose of "ensuring safety and security of personnel" and "mandated POC tasks"[25] when in

[20] Security Council Report, "The Financing of AU Peace Support Operations: Prospects for Progress in the Security Council?", April 26, 2023.

[21] Richard Gowan, "Planning Ahead: How the US May Recover Its Diplomatic Standing at the UN After the Gaza War," Just Security, 30 January 2024.

[22] https://dppa.un.org/sites/default/files/2019_update_to_the_multi-year_appeal.pdf

[23] https://www.un.org/peacebuilding/fr/fund

[24] United Nations Secretariat, Administrative Instruction, Gratis Personnel*, ST/AI/1999/6, 28 May 1999.

[25] C-34 Report, A/74/19, Mai 2020.

fact the UN Secretariat had developed a much more ambitious peacekeeping intelligence policy over the previous years, using XB funding. Funding the "Third UN" of *think tanks* and other NGOs around the UN can also be useful a way to advance certain agendas indirectly, through backdoor processes that are often more informal than heavy UN intergovernmental processes.

Although traditionally the remit of smaller European countries, other countries including China, which continues to claim to be an "emerging country" and to cultivate its special relationship with the countries of the South, known as the "non-aligned movement" or "G77" (Group of 132 developing countries in the General Assembly)[26] started adopting similar influence strategies in recent years. In 2016, China set up a "United Nations Trust Fund for Peace and Development"[27] worth $1 billion over ten years managed by the UN (with a Steering Committee composed of the Under-Secretary-General for Economic and Social Affairs from China, the Chief of Staff to Secretary-General Guterres, and Chinese diplomats). The fund is used to finance various activities including some related to peacekeeping agendas China is championing such as that of the safety and security of peacekeepers.[28] China also started financing Junior Professional Officers (JPO) positions—as well as supporting many internships—for young Chinese nationals to serve in the UN Secretariat—where China is underrepresented compared to many European countries notably who have long financed such JPO programs.

Diverging Interpretation of Principles, Norms and Values

Another set of consequences of these renewed power rivalries concerns the very basic principles that guide UN peacekeeping—consent of the parties and in particular of the host State; impartiality, and non-use of force except in self-defense or in defense of the mandate—as well as some of the norms and values that have traditionally guided the work of the organization. Although these principles are constantly recalled at the beginning of Council resolutions, they have been under strain. "Norms" are generally

[26] United Nations, "The Historical Significance Of The g77 | United Nations."

[27] https://www.un.org/fr/unpdf/index.shtml

[28] Kevin S. Kennedy and Laura Powers, "Senior Leadership Training in UN Peace Operations," International Peace Institute, February 2019.

defined as standards of behavior that are appropriate and accepted by actors with a given identity. In the context of peace operations, norms referred to by member states include for instance the respect for sovereignty, protection of civilians, human rights, gender, etc. Norms and values are also linked since norms are often built on shared values; even if the notion of "values" is more linked to ethics and morality in international relations.[29] This is also why the institutionalization of certain norms over time does not prevent them from being challenged someday, in particular if the context and the balance of power between blocks of states with like-minded identities and value systems evolve, which is what we are witnessing now.

Normative changes at the UN and in peacekeeping in particular are the result of complex processes, of power relations between Member States of course, but also involving think tanks and other NGOs themselves more or less influenced by the member states that fund them, what John Karlsrud calls "linked ecologies" between all these actors who form formal and informal alliances.[30] In the same vein, Tatiana Carayannis and Thomas G. Weiss in their book "*The Third United Nations*" describe how this "Third UN" in which they include a very broad group of non-state actors—intellectuals, academics, consultants, think tanks, NGOs, private sector and media—interacts with the intergovernmental mechanisms of the "First UN" (member states) and "the second UN" (the UN Secretariat or bureaucracy) to formulate and refine ideas and decision-making at key moments. Some champion particular ideas, others help analyze or operationalize their implementation; many thus help the UN to "think."[31]

The bigger powers, with plethoric teams of diplomats in their permanent representations to the UN in New York and in their capitals, but also with experience and know-how built up over time, generally rely very little on this "Third UN." This is what Vincent Pouliot calls "diplomatic power."[32] Conversely, the middle powers with significant financial means, such as the Nordic countries or Germany, have made the "Third UN" an important instrument of influence at the UN, including in support of their campaigns for non-permanent seats on the Security Council, to gain

[29] Jean-Baptiste Jeangène Vilmer ed., *Ethics of international relations. Contemporary issues*, Presses Universitaires de France, 2013, pp. 1–38.

[30] John Karlsrud, *Norm changes in international relations.*

[31] Tatiana Carayannis and G. Weiss, *The "Third One".*

[32] Vincent Pouliot, *International Pecking Orders: The Politics and Practice of Multilateral Diplomacy*. p. 131.

visibility but also to advance their ideas and agendas. Elected members of the Council have notably institutionalized the participation of civil society representatives as *briefers* in thematic sessions but also increasingly in formal Council sessions dealing with peace operations.[33]

This is how during the quarterly Council session on Mali in January 2023, a Malian human rights defender was invited by one of the non-permanent members of the Council as a civil society briefer. This is turn triggered the February 5, 2023 decision by the Malian government to expel the director of the MINUSMA's human rights division, accusing the UN of "politicizing and instrumentalizing the human rights issue," an allegation that Russia and China have echoed in the Council and which likely played an important part in the Malian decision a few month later to ask for the MINUSMA withdrawal altogether.[34] This is a worrying trend of certain states challenging norms such as human rights which they see as "Western liberal norms"—especially if they serve to criticize their own military and allies—even if these have guided peacekeeping since the end of the Cold War.[35] China and Russia justified their abstention on UNMISS's 2022 mandate, by stating that it read like "a Human Rights Council document" and the following year both abstained again on UNMISS's mandate based on the "very broad freedom of action" the mandate called for on the protection of civilians.[36] Chinese diplomats have actually tried in recent years to cut budgets for human rights components of peace operations under the banner of respect for sovereignty, but without much success so far.[37] In the past, Russia has also threatened to cut human rights and environmental protection position in the fifth committee if aviation budgets—which Russian contractors greatly profit from—were reduced.[38]

It is important to emphasize the role played by "the second UN" and the fact that the United Nations Secretariat in New York and also the field

[33] See for example this webinar from The Nordic African Institute "Civil Society Briefers in the UNSC".

[34] Security Council, S/PV.9012, April 7, 2022.

[35] Tarak Barkawi et Mark Laffey, "The Imperial Peace:: Democracy, Force and Globalization", *European Journal of International Relations* 5, nº 4 (Decembre 1999): 403–34, https://doi.org/10.1177/1354066199005004001.

[36] UN Security Council, 8994th Meeting, S/PV.8994, March 15, 2022; UN Security Council, 9281st Meeting, S/PV.9281, March 15, 2023.

[37] Louis Charbonneau, "China Pushes to Cut UN Human Rights Posts," Human Rights Watch, June 7, 2017.

[38] Jake Sherman, "With Peacekeeping Budget Approved, More Contentious Negotiations Lie Ahead", The Global Observatory, July 13, 2018.

missions are endowed with considerable room for maneuver,[39] which they cherish. Bureaucrats are in practice far from being simple executors of the decisions of the Security Council or the General Assembly, even if these bodies define the mandates of operations and control their budgets. This is particularly true for the issues of principles, norms and values, where bureaucratic resistance and/or inertia can counter member states influence. That said, the UN staff is far from being a monolithic block and it is not uncommon to see in the UN Secretariat as in field missions, divided staff or even sometimes "clans" themselves with diverging views or interpretations of the mandate and in particular the support that can/must or cannot/must not be given by the mission to the host State and civilian populations.

The basic principles of peacekeeping have been tested in recent years in contexts where there is little or no peace to keep, and where peacekeepers have to manage conflicts rather than implement peace agreements. In these debates on principles, different camps of Member States have emerged. While some have advocated for more "robust" peacekeeping, ready to use force against armed groups in particular, others have called for a return to more traditional maintenance of peace, observing ceasefire between parties without intruding too much on host state governance issues in particular, arguing that peacekeeping has become overloaded with tasks. India, for example, constantly warns against any ambiguity in the formulation of mandates which could cause peace operations to slide toward counterterrorism or simply offensive postures and would put the peacekeepers in danger.[40]

Bringing the Battle for Influence Out to Field Operations

While the most visible consequences of the return of power rivalries at the UN may be in member states discourses and postures in the Security Council in New York, consequences are also felt in field operations on the ground. Power rivalries are revealed through the behavior of the troop-contributing countries (TCCs), which are an increasingly different group

[39] Graham T. Allison & Morton H. Halperin. "Bureaucratic Politics: A Paradigm and Some Policy Implications." World Politics 24 (1972): 40–79. https://doi.org/10.2307/2010559.

[40] Security Council, S/PV.9082, June 29, 2022.

of countries from those Council members that mandate and rich countries who finance these operations. Troop contributing countries can notably resist the implementation of a mandate they are not happy with (or the interpretation of which they are not comfortable with), including by imposing operational limitations and other *caveats* during operations.

The deployment of forces in "parallel" to UN peace operations, often has a major influence on the conduct of peace operations, all the more if the same country who has a parallel force on the ground also places military advisers within the UN mission in key planning and operational functions. For instance, rather than deploying French contingents under blue helmets (as it does in Lebanon with UNIFIL), France seems to favor the deployment of French or European forces in "parallel" under their own national chain of command and rules of engagement. From France's point of view, this allows its troops to support peacekeepers when needed (in extremis support) while retaining the flexibility and robustness of a of a national military operation.[41]

More controversially, Russia has also been experimenting with the use of "parallel forces" to project influence and contest the influence of France in Central African Republic and Mali, but of a very different nature. The impact of the presence of private military contractor *Wagner Group* on UN peace operations is complex but has created additional challenges for UN peace operations, both operationally (in terms of safety and security,[42] including deconflicting airspace for instance, and obstructing human rights investigations[43]) but also and probably first and foremost in terms of managing misinformation and disinformation and the reputation of the UN.[44] While some parallel forces of the African Union notably may be complementary to or even enabling the work of the UN as in Somalia, others born out of bilateral agreements (the DRC inviting Ugandan soldiers in the east of the country at the start of 2022, or *Wagner Group* deployments in Central African Republic and Mali) are clear competitors that contribute to constraining and discrediting the action of blue helmets.

[41] Alexandra Novosseloff and Lisa Sharland, "Partners and Competitors: Forces Operating in Parallel to UN Peace Operations," November 2019.

[42] "Central African Republic: Four French Soldiers Arrested At Bangui Airport", France 24, 2022.

[43] RFI, "In the Central African Republic, victims of Russian exactions break the law of silence", 3 May 2021.

[44] https://minusca.unmissions.org/guira-fm

Another key actor in power games at field level is of course the host State. Although generally weak because in crisis at the time a peace operation is sent to their country, host states have become more assertive. Because without their consent, an operation cannot function. This is what we call the "power of the weak." Some host States maintain special relations with certain members of the Council and regional powers. They can also constrain the freedom of action of a UN operation and therefore the implementation of the mandate of the Security Council. In extreme cases, the host State can simply withdraw its consent to the presence of an operation on its territory as Mali did on June 16, 2023, or simply reject the appointment of a head of mission, boycott him, and/or demand the departure of a UN official from the country. In the current geopolitical context, host governments feel bolstered by the political support of the "pro-sovereignty" members of the P5, China and Russia, who no longer hesitate to abstain from voting on a resolution on a peace operation if they feel that the concerns of the host state have not been adequately addressed. Host states, from Colombia to Sudan, have also never had as great a say in the design and the conduct of peace operations, when in more and more cases UN peacekeepers are only one security offer among others they can chose from.

Finally, the special representatives and envoys of the UN Secretary-General and other commanders of UN forces on the ground (the *leadership* of the peace operation), although they have an important margin of maneuver within the perimeter of the mandate given by the Council, which they interpret, are not sheltered from the influence of powers either. The *leadership* of a peace operation is certainly a key factor in the success or failure of an operation. And the power rivalries particularly between P5 but also regional powers have led to a growing politicization of senior appointments.[45] While this is not a new issue, over the past few years, an increasing proportion of UN heads of mission, have been either imposed by host governments or major powers, or blocked by members of the P5 (and their allies). This compounded with the growing assertiveness of host countries in turn raises questions about the ability of peace operations to uphold the principle of impartiality.[46]

[45] We define politicization as the process of socialization by which an individual or an organization is led to take an interest in politics and to develop reflections and practices relating to it.

[46] Jeffrey Feltman, "Restoring (some) impartiality to UN senior appointments," Brookings, 29 October 2020.

Implications for the Future of Peace Operations

The analysis of peace operations cannot of course be reduced to that of geostrategic rivalries, but one cannot understand the evolution of UN peace operations since 1948 and *a fortiori* of the past twenty years without taking these rivalries into account, especially at a time when they these have growing (and mostly negative) consequences for peace operations. This conclusion militates for reconciling the study of UN peace operations which has been dominated by scholars of liberal international relations theory, with a more realist or realist-constructivist approach that better factor in the impact of the return of—primarily State—power rivalries onto peace operations.

Power dynamics at headquarters in New York and in the field are not the same but influence each other. What is decided at the Security Council in New York is not always followed by effects on the ground, but power rivalries in the field directly impact peace operations and inevitably end up affecting the dynamics of the Council and beyond as we have seen in the case of Mali, where the arrival of the *Wagner Group* first led France—also the pen holder on Mali in the Council—to take the decision to withdraw its parallel military force *Barkhane* in 2022, but not long after led the Malian transition government to ask for the departure of peacekeepers. While the commitment and support of Member States is essential in New York (political and financial support and contributions in troops), on the ground, too obvious interference from one or more powers, including in the information field, can have a brutal and lasting effect on a peace operation.

The UN and its peace operations not only reflect or reproduce inequalities between states, but they can also sometimes modify them and this is also why States attempt to shape and/or instrumentalize the organization to serve their interests. Middle powers and smaller member states can, for example, resist the will of most powerful through forging alliances, in the Security Council and in the General Assembly, but also in the field. Power and influence strategies can also be constrained by logics of resistance or simply bureaucratic inertia of the UN Secretariat and its field operations. The multiplicity of actors and the complexity of the mechanisms of power, influence and resistance within the multilateral system make it difficult for even the more powerful states to ultimately impose their will.

Some States seem to have realize the potential of the UN and of multilateralism as a means of influence and potentially building legitimacy. This is the case of China but also of certain liberal middle powers. Other states however, including some permanent members of the Security Council, but not only, seem to take their influence within the organization for granted, even when the legitimacy and authority of institutions like the Security Council, and their individual power and influence derived from their P5 status for instance, are increasingly being challenged. Other states also opt for a more "à la carte multilateralism", sometimes choosing to act through the UN when it can further their interests, but also sometimes prefer to act outside of the multilateral system when it best serves their interest. Interestingly, power and influence is often approached by states as an end in itself on the world stage rather than as a means of achieving strategic objectives that are not always well defined.

In view of the above analysis, we identify two main risks for the future of UN peace operations. The first risk is that the current cycle of heightened power rivalries continues to constrain UN operations and their effectiveness, and to undermine their relevance with the risk that it becomes as self-fulfilling prophecy. We observe that such situations of rivalries sometimes produce deadlocks over mandates and senior appointments notably. But the second risk is that on the contrary power rivalries lead to great and middle powers becoming completely disinterested and therefore disengaging durably from peace operations that they see as neither delivering on the ground, nor really serving their interests.

While it is still early to assess the long-term implications of the invasion of Ukraine, it is urgent to ensure that the divisions within the P5 do not lead to a lasting disengagement from the multilateral system. A likely scenario is that heightened power rivalries in certain regions may lead to fewer peace operations there because the P5 and their proxies can't find a mutually advantageous compromise mandate and operation, but that in other regions, small windows may remain for tailored operations, not necessarily a UN peacekeeping force but an operation that would involve some kind of UN support. This could for instance be the case in Haiti. While it was unconceivable until two years ago that the US would envisage a new multilateral operation there and that the host country would accept a return of the UN (after MINUSTAH brought cholera to the country notably), the deteriorating security situation and absence of credible alternative non-UN force may even ultimately lead to a UN force or at least a support package.[47] In Africa, the future of UN operations may be in smaller UN outfits in support of African operations which would better respond to the expectations of host states for offensive/fighting forces but don't have the logistical and financial capacity, and lack some key civilian substantive functions.

Contrary to what has often been written in the academic and policy literature,[48] the hardening of international relations has not necessarily resulted in more robust or offensive operations. Despite brief UN experiments such as the Force Intervention Brigade (FIB) in eastern Democratic Republic of Congo, offensive operations remain for the most part the prerogative of few capable national forces, NATO and some regional coalitions. But power rivalries have led a number of UN operations to either attempt to marginalize (Taliban in Afghanistan, Hezbollah in southern Lebanon, etc.) or target armed groups (gangs in Haiti, M23 in eastern DRC, *Seleka* in the Central African Republic, etc.). In many of these cases, the UN has been increasingly perceived by actors on the ground as being part of Western stabilization efforts and acting for the interests of certain powers and/or their allies in the region, thereby compromising its basic principle of impartiality.

In this moment of heightened geopolitical rivalries, many peace operations don't really have a choice but to learn to live with a divided Security Council, and to try to manage the various influence strategies of member states, while trying to preserve a certain impartiality without falling into figuration or submitting to all the injunctions of the host State and risking becoming a service provider for governments which often lack local legitimacy in at least part of the country. It is however essential that permanent members of the Security Council and other powerful states do not unduly

[47] Eugene Chen, "How Can the UN Best Support Kenya's Mission to Haiti? Delay Provides Time to Reflect," The Global Observatory, January 31, 2024.

[48] See for instance Robert Muggah, "The United Nations Turns to Stabilization", IPI Global Observatory, 5 December 2014; Cedric de Coning, *UN Peacekeeping Doctrine in a New Era: Adapting to Stabilisation, Protection and New Threats*, 2017, Routledge; Aditi Gorur, *Defining the Boundaries of UN Stabilization Missions*, Washington, DC: Stimson Centre, 2016; John Karlsrud, "The UN at War: Examining the Consequences of Peace-Enforcement Mandates for the UN Peacekeeping Operations in the CAR, the DRC and Mali." *Third World Quarterly* 36, no. 1, 2 January 2015; John Karlsrud, "Towards UN Counter-Terrorism Operations?" *Third World Quarterly* 38, no. 6, June 3, 2017: 1215–31; Charles T. Hunt, "All Necessary Means to What Ends? The Unintended Consequences of the 'Robust Turn' in UN Peace Operations", *International Peacekeeping* 24, n° 1, January 2017, 108–31.

politicize multilateral institutions in general and do not sacrifice UN peace operations on the altar of their power rivalries. Great and middle powers must be able to use peace operations while serving them, but without compromising or doing them too much harm, because they would otherwise risk depriving themselves in the future of this important tool of multilateralism.

China's growing assertiveness and its sometimes clumsy but diversified strategies of influence at the UN will likely be decisive for the future of peace operations. China's long-term strategy and its significant investment in the multilateral system seem to provide some protection for peace operations against possible Russian "revisionist" tendencies in the short term. We have for instance witnessed this at the occasion of the renewal of peace operations that followed the February 2022 invasion of Ukraine. China would have played a discrete but effective lobbying role with its Russian colleagues so that they wouldn't be tempted to retaliate against western powers by vetoing the renewal of certain peace operations. But China also seeks to use and shape international institutions to its advantage and this could have a major impact on peace operations principles, norms and values in the medium term.

In this context, the continued and active engagement of "constructive" middle and great powers will be key in trying to preserve peace operations as a key tool of multilateralism, while trying to reform it so that it is more relevant to current conflict dynamics, but without compromising on some important norms and values such as human rights. The room for maneuver vis-à-vis great powers is of course often limited. And the inertia and sometimes bureaucratic resistance in the field of norms and values in the current context can be a blessing. The "third UN" of non-state actors, *think tanks* and NGOs, also plays an increasingly significant role in these games of influence, particularly in the domain of ideas.

The UN faces a dilemma as to its level of ambition. On the one hand, peace operations are tempted to constantly adapt to the new challenges and conflicts they face in order to continue to interest and "be useful" and relevant to Member States. On the other hand, the current context considerably reduces their room for maneuver, notably because of the lack of consensus in the Security Council around mandate renewals, but also the deep disagreements on the political strategies to be pursued and the norms and values that missions must uphold and promote.

A "back to basics" with smaller modular and adaptable type of missions—which could act as a "hub" of conflict management expertise in tandem with other actors as some have suggested[49]—could help maintain a minimum consensus among Member States around the value of the tool. It would, however, risk producing missions without political and military levers confined to a relatively passive and powerless observer role, and vulnerable to deterioration in the security situation as seen with UNITAMS which had to leave Khartoum in 2022 following the resumption of fighting or UNSMIL which had to leave Tripoli for Tunis, or the short-lived UNSMIS in Syria, years before that. While "low key" missions can sometimes safeguard space for the UN Secretariat to operate below the radar of powerful states, the risk is that this same states become disinterested in the work of the UN and turn to other channels to project their power and influence, through existing or new bilateral agreements, the deployment of mercenaries, or more generally the use of force too often at the expense of more political solutions.

[49] Richard Gowan and Louise Riis Andersen, What Next for UN Peace Operations? Copenhagen, Danish Institute for International Studies, June 12, 2020.

CHAPTER 7

Conclusion: A New Agenda for Peace Operations

"My clear opinion today is that peacekeeping operations, where there is no peace to keep, do not really make much sense and I am interested in finding orderly exit strategies in order to be able to replace those peacekeeping forces by other forms, more effective, to do the job."[1]

Secretary-General António Guterres, 22 June 2023

Reports of the death of United Nations Peace operations have been greatly exaggerated. When, after securing a second five-year term in 2021, UN Secretary-General António Guterres released his vision for the future of global cooperation in "Our Common Agenda," many of those working in or on UN peace operations grew concerned about the report's virtual omission of the topic. The only passing reference to peacekeeping related to concerns of "liquidity reserves" reflecting circumstantial cash-flow problems and financial strains that had impacted missions and the ability to reimburse troop-contributing countries and continue to affect the regular budget of the organization. The latest UN reform agenda released in 2023 by Secretary-General António Guterres, "The New Agenda for Peace" (NA4P), did however give some recognition to peace operations, which it describes as "an essential part of the diplomatic toolbox" of the

[1] António Guterres, Lecture and conversation with students at Sciences Po (Paris School of International Affairs), June 22, 2023.

A. Boutellis, *The UN Stabilization Mission in Mali (MINUSMA)*, Twenty-first Century Perspectives on War, Peace, and Human Conflict, https://doi.org/10.1007/978-3-031-66466-3_7

UN and suggests that "these missions will remain a central component of the continuum of United Nations responses" as "peacekeeping represents effective multilateralism in action."[2] Some similar language is included in the *Pact for the Future* that was endorsed by Heads of State/Government at the September 2024 Summit after a long negotiation to the wire led by co-facilitators Germany and Namibia, including an attempt by Russia to introduce last minute changes. While the implementation of the *Pact* may very well frame much of the discussion on peace operations for years to come, this will not guarantee the future of these operations.

The major peace operations trend of the past two decades has been the end of major multidimensional peacekeeping operations in Africa. These missions, which range from 13,000 to more than 20,000 troops, police, and civilian personnel each, have often struggled to manage conflicts rather than keep peace, leading to what the NA4P describes as "the gap between United Nations peacekeeping mandates and what such missions can actually deliver in practice." Although post-Cold War Western liberal interventionism has primarily been exemplified by the failed experiences of the US/NATO in Afghanistan and Iraq and to a lesser degree France/Europe in the Sahel, these types of large UN operations—some of which are termed "stabilization" without a consensus on what the term encompasses—have also sometimes become seen through the same interventionism prism, particularly when deployed in parallel or on the heels of military interventions. It is undeniable that UN peace operations are facing a crisis of relevance in the face of both the changing nature of conflicts (fragmentation and internationalization) and changing geopolitics. And the NA4P calls for the organization to "adapt to the geopolitical realities of today and the threats of tomorrow."

In many ways, the NA4P confirmed the vision Guterres has laid out since arriving at the helm of the World organization in 2017. Earlier this summer, when asked by a student whether there is a future for UN peacekeeping, Guterres answered, "My clear opinion today is that peacekeeping operations, where there is no peace to keep, do not really make much sense and I am interested in finding orderly exit strategies in order to be able to replace those peacekeeping forces by other forms, more effective, to do the job." Transition and exit strategies are indeed at the core of the NA4P, suggesting that the secretary-general wants to move on from

[2] Secretary-General, "Our Common Agenda Policy Brief 9: A New Agenda for Peace," July 2023.

current peace operations, which were—Mali and Sudan—or are—Democratic Republic of the Congo—at an impasse. This also confirms a broader trend of contraction, with the closing of other major operations in 2017–2018 in Liberia, Côte d'Ivoire, and Haiti, followed by the end of the hybrid UN-AU operation in Darfur at the end of 2020. The NA4P reiterates calls on member states to ensure the "primacy of politics," and recognizes a "normative challenge" and "significant global retrenchment of human rights" but also openly embraces "multinational peace enforcement and counterterrorism and counter-insurgency operations," including enforcement action by regional and sub-regional organizations. In many ways, this is the direction most member states affected by terrorism have already taken, often prioritizing bilateral—in some cases *Wagner Group* mercenaries—or ad hoc coalitions over multilateral arrangements.

In a context of competition for influence, especially on the African continent, more rulers prefer military support that is quicker to deploy and more willing to respond to urgent security needs, without the human rights and other political governance strings attached and oversight from a Security Council whose authority they are no longer afraid to challenge. Despite the passing of The framework Resolution 2719 (2023) on the financing of AU-led peace support operations (AUPSOs) on December 21, 2023, ultimately, decisions to deploy such non-UN operations under Chapter VII of the UN Charter will continue to be case-by-case and interest-based. Contentious accountability and regional AU and sub-regional (RECs) leadership issues—on display in the military coups across West Africa and the Tigray/Ethiopia and Sudan crises—along with operational and political challenges already experienced by partnership peacekeeping under AMISOM will not magically go away.

It is difficult to make predictions, especially about the future of peace operations. Secretary-General Boutros-Ghali's January 1992 Agenda for Peace—which was published at a time when Cold War constraints had been lifted and peacekeeping operations proliferated (22 new operations were launched between 1988 and 1993!) as a response to interstate and regional conflicts—failed to anticipate the rapid evolution toward "multidimensional" peacekeeping as a response to suddenly dealing with civil wars. One thing the study of the world organization teaches us though is that since its creation in 1948, peace operations have adapted and been shaped by developments in the international order. They have also often be used as a last resort, when everything else fails. One should therefore be mindful not to bet against the future and underestimate the resiliency of UN peace operations.

Interestingly, the *Pact for the Future* requests the Secretary-General to undertake a review on the future of all forms of United Nations peace operations, taking into account lessons learned from previous and ongoing reform processes, when the NA4P called on member states to "undertake a reflection on the limits and future of peacekeeping." This could be a risky exercise on different levels. Member states, in addition to being overwhelmed by Guterres's many policy briefs and initiatives, could feel a sense of fatigue: there was the 2015 Report of the Independent High-level Panel on Peace (HIPPO), followed by the 2017 Cruz report, the 2018 Action for Peacekeeping (A4P) that led to a negotiated Declaration of Shared Commitments on UN Peacekeeping Operations—endorsed by more than 150 member states—and the 2021 A4P+, as well as various related resolutions. Another risk is that reopening this debate in the current geopolitical context could weaken rather than strengthen a fragile consensus around peace operations, at a time when member states' unity continues to erode as illustrated by more difficult negotiation processes and a growing number of abstentions on peacekeeping resolutions since 2018. At the end of the day, this may turn into a synthesis exercise that think tanks and academics would be well placed to support, even if it concludes in a repeat of past lessons already encapsulated in the 2015 HIPPO report, the 2008 Capstone Doctrine, and the 2000 Brahimi Report, including that the Secretariat should tell the Security Council what it needs to know, not what it wants to hear.

One of the fundamental questions for peace operations in the years to come will be that of principles and norms. Member states, including host governments, have increasingly challenged some of the basic principles that have guided peacekeeping—consent of the parties, impartiality, and non-use of force except in self-defense and defense of the mandate. These challenges have been exacerbated by the Security Council mandating peacekeeping operations to support the restoration or extension of dysfunctional states that often lack legitimacy in the eyes of local populations. Mandates related to human rights and gender have also been challenged by some as encroaching on sovereignty and promoting Western agendas.

But many member states across different regional groupings, and despite their power rivalries and divergent interests and values increasingly affecting peace operations, continue to invest considerable political, financial, and human resources in peacekeeping—not least China, which is now the second-highest contributor to the UN peacekeeping budget behind the United States. Despite the many challenges, they still see peace operations as one of the main peace and security tools at the disposal of the Security Council and an important vehicle through which they can practice multilateralism. And even if the 2022 invasion of Ukraine and the return of interstate conflicts are the subject of much attention, one must not lose sight of the fact that many other smaller scale inter and intrastate conflicts continue and will continue to exist. For these, and where power rivalries are perhaps less intense, the traditional tools of the UN—prevention, mediation and peacekeeping—will remain relevant even if they will necessarily have to be adapted. If the idea behind the NA4P was to throw the ball back to member states, more reflection will be needed for member states to chart what future peace operations could be in a time of renewed geopolitical rivalries, including one-dimensional and "nimble adaptable models" as well as creative partnership models. While some cynics may think UN peace operations could benefit from less member state attention these days, the fact is that a total disengagement from great and middle powers would be even more detrimental not only to peace operations but also to the UN organization more broadly.

This book demonstrates how UN peace operations, which had long been the subject of a consensus among Member States and were sheltered from power rivalries because they helped manage peripheral crises, have become another field of confrontation for the great and middle powers with increasingly divergent interests and values, both at headquarters in New York and at the level of field operations. These operations are both instrument and object of these new power dynamics. Although Security Council members had demonstrated a significant ability to "compartmentalize" their differences, and thus maintain a common front on the majority of peace operations, growing rivalries since 2018 started harming these operations and their effectiveness.[3] However, a total disinterest and disengagement of these powerful states would be even more detrimental to peace operations at a time of a bigger crisis of multilateralism.[4]

The current geostrategic environment means that the United Nations and its peace operations can no longer hide behind the vague umbrella of

[3] Lise Morjé Howard, *UN Peacekeeping in Civil Wars*, 1st ed. (Cambridge University Press, 2007), https://doi.org/10.1017/CBO9780511840593.

[4] Julian Fernandez and Jean-Vincent Holliner (eds.), *DISUNITED NATIONS?: The crisis of multilateralism in international relations*, Publisher: CNRS editions, 2022. p. 20.

the so-called international community or take for granted unanimous votes of peacekeeping resolutions by a Security Council that is increasingly divided. Rather than try to shelter itself and risk irrelevance, the UN Secretariat should embrace power rivalries and better consider power dynamics at headquarters and in field operations without resigning itself to them. As Raymond Aron writes, we must strive "to keep a cool head, to observe the world as it is, and not to imagine it as we would like it to be."[5]

For the UN Secretariat, the question then arises of how to preserve and nurture the interest of the powers in the peace operations tool, and thereby guarantee its political and financial sustainability, while also keeping powerful states sufficiently at a distance to limit excessive interference in the conduct of these operations. The challenge for the organization is to be truly geopolitical while maintaining a certain impartiality and legitimacy, and to preserve a delicate balance between the UN of member states and the UN of "We, the peoples of the United Nations" as per the UN Charter, the UN of the few great powers including those of the P5, and the UN of all member states.

For their part, member states would be mistaken to neglect multilateralism in the context of the hardening of international relations, or to focus exclusively on bilateral and ad hoc peace enforcement arrangements and "minilateral" coalitions, instead of continuing to develop new models as part of a spectrum of UN peace operations that can respond to the evolving nature of conflicts. Beyond the vision centered on rank, today's world requires diversified strategies of power and influence (*soft power*, *balancing*, indirect influence, etc.) for which multilateralism can actually be a useful vehicle. This is all the more important for Western powers in an increasingly "non-Western" world, with new influential actors from the "Global South" and in which "non-alignment"—or "multi-alignment" rather[6]—is making a comeback in response to the re-emergence of an East-West divide.

[5] Raymond Aron, "In Search of a Foreign Policy Philosophy." p. 82.

[6] Shreya Upadhyay, "BRICS, Quad, and India's Multi-Alignment Strategy", Stimson Center, July 12, 2022.

References

Allison, G. T., & Halperin, M. H. (1972). Bureaucratic Politics: A Paradigm and Some Policy Implications. *World Politics, 24*(S1), 40–79. https://doi.org/10.2307/2010559

Aron, R. (1953). En quête d'une philosophie de la politique étrangère. *Revue française de science politique, 3*(1), 69–91. https://doi.org/10.3406/rfsp.1953.452691

Aron, R. (2004). *Paix et guerre entre les nations avec une présentation de l'auteur.* Calmann-Lévy.

Autesserre, S. (2011). Construire la paix: conceptions collectives de son établissement, de son maintien et de sa consolidation. *Critique internationale, 51*(2), 153. https://doi.org/10.3917/crii.051.0153

Autesserre, S. (2014). *Peaceland: Conflict Resolution and the Everyday Politics of International Intervention.* Cambridge University Press.

Autesserre, S. (n.d.). *The Crisis of Peacekeeping. Why the UN Can't End Wars.* Retrieved from https://www.foreignaffairs.com/articles/2018-12-11/crisis-peacekeeping

Badie, B. (2007). Le Défi de La Puissance. In *Le Multilatéralisme* (pp. 215–232). La Découverte. https://doi.org/10.3917/dec.badie.2007.01.0215

Badie, B., & Devin, G. (Eds.). (2007). *Le Multilatéralisme: Nouvelles Formes de l'action Internationale. Textes à l'appui.* Découverte.

Badie, B., Vidal, D., & Amiel, F. (2021). La France, une puissance contrariée: l'état du monde 2022.

A. Boutellis, *The UN Stabilization Mission in Mali (MINUSMA)*, Twenty-first Century Perspectives on War, Peace, and Human Conflict, https://doi.org/10.1007/978-3-031-66466-3

Ballou, B. (2014). Why America's Nation Building Office Failed and What Congress Had to Do With It. *Stability: International Journal of Security & Development, 3*(1). https://doi.org/10.5334/sta.ec
Barkawi, T., & Laffey, M. (1999). The Imperial Peace:: Democracy, Force and Globalization. *European Journal of International Relations, 5*(4), 403–434. https://doi.org/10.1177/1354066199005004001
Barkin, J. S. (2012). *Realist Constructivism Rethinking International Relations Theory*. Cambridge University Press. Retrieved from http://0-www.ebooks.cambridge.org.cataleg.uoc.edu/ebook.jsf?bid=CBO9780511750410
Barnett, M., & Duvall, R. (2005). Power in International Politics. *International Organization, 59*(01). https://doi.org/10.1017/S0020818305050010
Barnett, M. N., & Finnemore, M. (1999). The Politics, Power, and Pathologies of International Organizations. *International Organization, 53*(4), 699–732. https://doi.org/10.1162/002081899551048
Battistella, D. (2015). *Théories des relations internationales. 5e édition mise à jour*. Sciences PoLes Presses.
Battistella, D., Cornut, J., & Baranets, E. (2019). Théories des relations internationales.
Baumann, A. B. (2008). Clash of Organisational Cultures? The Challenge of Integrating Civilian and Military Efforts in Stabilisation Operations. *The RUSI Journal, 153*(6), 70–73. https://doi.org/10.1080/03071840802670122
Bellamy, A. J., & Hunt, C. T. (2015). Twenty-First Century UN Peace Operations: Protection, Force and the Changing Security Environment. *International Affairs, 91*(6), 1277–1298. https://doi.org/10.1111/1468-2346.12456
Bellamy, A. J., & Williams, P. D. (Eds.). (2013). *Providing Peacekeepers: The Politics, Challenges, and Future of United Nations Peacekeeping Contributions*. Oxford University Press.
Bellamy, A. J., Williams, P. D., & Griffin, S. (2010). *Understanding Peacekeeping* (2nd ed.). Polity.
Berdal, M. (2018). The State of UN Peacekeeping: Lessons from Congo. *Journal of Strategic Studies, 41*(5), 721–750. https://doi.org/10.1080/01402390.2016.1215307
Bode, I. (2020). Use of Force in Peace Operations. In *The Palgrave Encyclopedia of Peace and Conflict Studies* (pp. 1–8). Springer International Publishing. https://doi.org/10.1007/978-3-030-11795-5_83-1
Bogdan, R., & Taylor, S. J. (1975). *Introduction to Qualitative Research Methods: A Phenomenological Approach to the Social Sciences*. Wiley.
Boutellis, A. (2013a). Chad and the Central African Republic. In J. Boulden (Ed.), *Responding to Conflict in Africa* (pp. 121–144). Palgrave Macmillan US. https://doi.org/10.1057/9781137367587_7
Boutellis, A. (2013b). From Crisis to Reform: Peacekeeping Strategies for the Protection of Civilians in the Democratic Republic of the Congo. *Stability:*

International Journal of Security & Development, 2(3), 48. https://doi.org/10.5334/sta.ci

Boutellis, A. (2015). Can the UN Stabilize Mali? Towards a UN Stabilization Doctrine? *Stability: International Journal of Security & Development, 4*(1). https://doi.org/10.5334/sta.fz

Carayannis, T., & Weiss, T. G. (2021). *The "Third Un"* (1st ed.). Oxford University Press.

Center on International Cooperation. (2008). Annual Review of Global Peace Operations 2008. In United Nations (Ed.), *Peacekeeping Best Practises Unit.*

Center on International Cooperation. (2009). Annual Review of Global Peace Operations 2009. In United Nations (Ed.), *Peacekeeping Best Practises Unit.* Lynne Rienner Publishers.

Center on International Cooperation. (2010). Annual Review of Global Peace Operations 2010. In United Nations (Ed.), *Peacekeeping Best Practises Unit.* Lynne Rienner Publishers.

Center on International Cooperation, and United Nations. (2006). *Peacekeeping Best Practises Unit. Annual Review of Global Peace Operations 2006.* Lynne Rienner Publishers.

Chandler, D. (2010). *International Statebuilding: The Rise of Post-Liberal Governance. Critical Issues in Global Politics 2.* Routledge.

Charbonneau, B. (2017). Intervention in Mali: Building Peace between Peacekeeping and Counterterrorism. *Journal of Contemporary African Studies, 35*(4), 415–431. https://doi.org/10.1080/02589001.2017.1363383

Charillon, F. (Ed.). (2021). *La France Dans Le Monde. Coll. "Dans Le Monde".* CNRS éditions.

Charillon, F. (2022). Guerres d'influence: Les États à La Conquête Des Esprits. Odile Jacob.

Claude, I. L. (1963). United Nations Use of Military Force. *Journal of Conflict Resolution, 7*(2), 117–129. https://doi.org/10.1177/002200276300700203

Clayton, G., Dorussen, H., & Böhmelt, T. (2020). United Nations Peace Initiatives 1946–2015: Introducing a New Dataset. *International Interactions, 19*, 1–20. https://doi.org/10.1080/03050629.2020.1772254

Clunan, A. L. (2014). Why Status Matters in World Politics. In T. V. Paul, D. W. Larson, & W. C. Wohlforth (Eds.), *Status in World Politics* (pp. 273–296). Cambridge University Press. https://doi.org/10.1017/CBO9781107444409.017

Cold-Ravnkilde, S. M., & Jacobsen, K. L. (2020). Disentangling the Security Traffic Jam in the Sahel: Constitutive Effects of Contemporary Interventionism. *International Affairs, 96*(4), 855–874. https://doi.org/10.1093/ia/iiaa093

Cold-Ravnkilde, S. M., & Nissen, C. (2020). Schizophrenic Agendas in the EU's External Actions in Mali. *International Affairs, 96*(4), 935–953. https://doi.org/10.1093/ia/iiaa053

Coleman, K. P., & Job, B. L. (2021). How Africa and China Mai Shape UN Peacekeeping beyond the Liberal International Order. *International Affairs, 97*(5), 1451–1468. https://doi.org/10.1093/ia/iiab113

Coleman, K. P., & Williams, P. D. (2021). Peace Operations Are What States Make of Them: Why Future Evolution Is More Likely than Extinction. *Contemporary Security Policy, 8*, 1–15. https://doi.org/10.1080/13523260.2021.1882802

Constantin, C. (2000). *Par-delà l'équilibre: facteurs systémiques, motivations et attributs nationaux dans la formation des alliances.* Chaire Téléglobe Raoul-Dandurand.

Cooley, A., & Nexon, D. H. (2020). *Exit from Hegemony: The Unraveling of the American Global Order.* Oxford University Press.

Cox, R. W., & Jacobson, H. K. (1973). *The Anatomy of Influence: Decision Making in International Organization.* Yale University Press.

Crozier, M. (1971). *Le Phénomène bureaucratique: Essai sur les tendances bureaucratiques des systèmes d'organisation modernes et sur leurs relations en France avec le système social et culturel.* Points.

Cunliffe, P. (2012). Still the Spectre at the Feast: Comparisons between Peacekeeping and Imperialism in Peacekeeping Studies Today. *International Peacekeeping, 19*(4), 426–442. https://doi.org/10.1080/13533312.2012.709751

Cunliffe, P. (2013). *Legions of Peace: UN Peacekeepers from the Global South.* Hurst.

Curran, D., & Holtom, P. (2015). Resonating, Rejecting, Reinterpreting: Mapping the Stabilization Discourse in the United Nations Security Council, 2000–14. *Stability: International Journal of Security & Development, 4*(1), 50. https://doi.org/10.5334/sta.gm

da Costa, D. F., & Karlsrud, J. (2013). 'Bending the Rules': The Space between HQ Policy and Local Action in UN Civilian Peacekeeping. *Journal of International Peacekeeping, 17*(3–4), 293–312. https://doi.org/10.1163/18754112-1704007

Day, A. C., & Hunt, C. T. (2020). UN Stabilisation Operations and the Problem of Non-Linear Change: A Relational Approach to Intervening in Governance Ecosystems. *Stability: International Journal of Security and Development, 9*(1), 2. https://doi.org/10.5334/sta.727

de Coning, C. (2015). Towards More People-Centric Peace Operations: From 'Extension of State Authority' to 'Strengthening Inclusive State-Society Relations'. *Stability: International Journal of Security & Development, 4*(1), 49. https://doi.org/10.5334/sta.gl

de Coning, C. (2018). Adaptive Peacebuilding. *International Affairs, 94*(2), 301–317. https://doi.org/10.1093/ia/iix251

de Coning, C. (2019). How UN Peacekeeping Operations Can Adapt to a New Multipolar World Order. *International Peacekeeping, 26*(5), 536–539. https://doi.org/10.1080/13533312.2019.1677286

de Coning, C. (2020). Adaptive Peace Operations: Navigating the Complexity of Influencing Societal Change Without Causing Harm. *International Peacekeeping, 27*(5), 836–858. https://doi.org/10.1080/13533312.2020.1797500

de Coning, C. (2021). The Future of UN Peace Operations: Principled Adaptation through Phases of Contraction, Moderation, and Renewal. *Contemporary Security Policy, 5*, 1–14. https://doi.org/10.1080/13523260.2021.1894021

de Coning, C., & Peter, M. (2019). *United Nations Peace Operations in a Changing Global Order.* Retrieved from http://banques.enap.ca/Proxy.pl?adresse=https://doi.org/10.1007/978-3-319-99106-1

De Coning, C., Aoi, C., & Karlsrud, J. (Eds.). (2017a). *UN Peacekeeping Doctrine in a New Era: Adapting to Stabilisation, Protection and New Threats.*

De Coning, C., Aoi, C., & Karlsrud, J. (Eds.). (2017b). *UN Peacekeeping Doctrine in a New Era: Adapting to Stabilisation, Protection and New Threats. Global Institutions 125.* Routledge, Taylor & Francis Group.

de Montbrial, T. (2021). Rivalités de Puissance, Idéologies et Multilatéralisme. *Revue Défense Nationale, 838*(3), 18–23. https://doi.org/10.3917/rdna.838.0018

Deit, T. (n.d.). *Privatisation de La Paix: Réalité, Défis et Raisons.* Retrieved from https://www.academia.edu/34131730/Privatisation_de_la_paix_réalité_défis_et_raisons?email_work_card=title

Delon, F. (1993). La concertation entre les membres permanents du Conseil de Sécurité. *Annuaire français de droit international, 39*(1), 53–64. https://doi.org/10.3406/afdi.1993.3121

Devin, G. (Ed.). (2009). *Faire la paix: la part des institutions internationales. Références Mondes.* Presses Sciences Po.

Devin, G. (2013). Paroles de Diplomates: Comment Les Négociations Multilatérales Changent La Diplomatie. In *Négociations Internationales* (pp. 77–104). Presses de Sciences Po. https://doi.org/10.3917/scpo.placi.2013.01.0077

Devin, G. (2016). *Méthodes de recherche en relations internationales. Relations internationales 6.* SciencesPo les presses.

Devin, G. (n.d.). *Le Covid-19 et Le Multilatéralisme, Tribune.* Retrieved from https://www.la-croix.com/Debats/Forum-et-debats/Le-Covid-19-multilateralisme-2020-03-21-1201085310

Di Razza, N. (n.d.). *UN Peacekeeping and the Protection of Civilians in the COVID-19 Era.* Retrieved from https://theglobalobservatory.org/2020/05/un-peacekeeping-protection-of-civilians-in-covid-19-era/

Diehl, P. F., & Druckman, D. (2015). In J. A. Koops, T. Tardy, N. MacQueen, & P. D. Williams (Eds.), *Evaluating Peace Operations*. Oxford University Press. https://doi.org/10.1093/oxfordhb/9780199686049.013.3

Doyle, M. W., & Sambanis, N. (2006). *Making War and Building Peace: United Nations Peace Operations. Princeton Paperbacks*. Princeton University Press.

Durch, W. J., Holt, V. K., Earle, C. R., & Shanahan, M. K. (2003). *The Brahimi Report and the Future of UN Peace Operations*. Henry L. Stimson Center.

Dyer, W. G., & Wilkins, A. L. (1991). Better Stories, Not Better Constructs, To Generate Better Theory: A Rejoinder to Eisenhardt. *Academy of Management Review, 16*(3), 613–619. https://doi.org/10.5465/amr.1991.4279492

Eilstrup-Sangiovanni, M., & Hofmann, S. C. (2020). Of the Contemporary Global Order, Crisis, and Change. *Journal of European Public Policy, 27*(7), 1077–1089. https://doi.org/10.1080/13501763.2019.1678665

Fernandez, J., & Holeindre, J.-V. (2022). *NATIONS DÉSUNIES?: La crise du multilatéralisme dans les relations internationales*. CNRS éditions.

Fernandez, J., & Vilmer, J.-B. J. (Eds.). (2020). *Les Opérations Extérieures de La France. Biblis 229*. CNRS éditions.

Finnemore, M. (2005). Fights about Rules: The Role of Efficacy and Power in Changing Multilateralism. *Review of International Studies, 31*(S1), 187–206. https://doi.org/10.1017/S0260210505006856

Finnemore, M., & Sikking, K. (1998). International Norm Dynamics and Political Change. *International Organization, 52*(4), 887–917.

Fortna, V. P. (2008). *Does Peacekeeping Work? Shaping Belligerents' Choices after Civil War*. Princeton Univ. Press.

FP. (n.d.). *How Trump and Putin Weakened U.N. Bid for a Global Cease-Fire*. Retrieved from https://foreignpolicy.com/2020/04/17/trump-putin-global-ceasefire-un-coronavirus/?utm_source=PostUp&utm_medium=email&utm_campaign=20948&utm_term=Editors%20Picks%20OC&

Fröhlich, M., & Williams, A. (Eds.). (2018). *The UN Secretary-General and the Security Council* (Vol. 1). Oxford University Press. https://doi.org/10.1093/oso/9780198748915.001.0001

Galula, David. Counterinsurgency Warfare: Theory and Practice, 1968

Genevaz, J. (2015). La Chine et les opérations de maintien de la paix de l'ONU: défendre la souveraineté. *Politique étrangère, 80*(4).

Gilder, A. (2019). The Effect of 'Stabilization' in the Mandates and Practice of UN Peace Operations. *Netherlands International Law Review, 66*(1), 47–73. https://doi.org/10.1007/s40802-019-00128-4

Giuliani, C. (2021). Advancing Peace through Human Rights in UN Peace Operations: A Policy Perspective on Achievements and the Way Forward. *Journal of International Peacekeeping, 24*(1–2), 252–284. https://doi.org/10.1163/18754112-20210006

Gorur, A. (2020). *The Political Strategy of the UN Peacekeeping Mission in Mali.* UNU/Stimson Center.
Gorur, A. (n.d.). *Defining the Boundaries of UN Stabilization Missions.* Stimson Center.
Gowan, R. (2018). *The Price of Order.* Retrieved from https://cpr.unu.edu/the-price-of-order.html
Gowan. (n.d.). *Remembering Javier Pérez de Cuéllar's 'Piecemeal' Approach to UN Peacemaking.* Retrieved from https://www.crisisgroup.org/global/remembering-javier-perez-de-cuellars-piecemeal-approach-un-peacemaking
Gowan, R., & Stedman, S. J. (2018). The International Regime for Treating Civil War, 1988–2017. *Daedalus, 147*(1), 171–184. https://doi.org/10.1162/DAED_a_00482
Guéhenno, J.-M. (2015). *The Fog of Peace: A Memoir of International Peacekeeping in the 21st Century.* Brookings Institution Press.
Guéhenno, J.-M. (2018). The United Nations & Civil Wars. *Daedalus, 147*(1), 185–196. https://doi.org/10.1162/DAED_a_00483
Guichaoua, Y. (2020). The Bitter Harvest of French Interventionism in the Sahel. *International Affairs, 96*(4), 895–911. https://doi.org/10.1093/ia/iiaa094
Guterres, A. (2017). *Le Maintien de La Paix Est Rentable, Mais Doit s'adapter à La Nouvelle Réalité.* Retrieved from https://www.un.org/sg/fr/content/sg/articles/2017-05-30/peacekeeping-cost-effective-must-adapt-new-reality
Hatto, R. (2009). UN Command and Control Capabilities: Lessons from UNIFIL's Strategic Military Cell. *International Peacekeeping, 16*(2), 186–198. https://doi.org/10.1080/13533310802685489
Hatto, R. (2015). *Le maintien de la paix. L'ONU en action.* Armand Colin.
Hecker, M., & Tenenbaum, E. (2021). *La Guerre de Vingt Ans: Djihadisme et Contre-Terrorisme Au XXIe Siècle.* Robert Laffont.
HIPPO. (2015). *Rapport Du Groupe Indépendant de Haut Niveau Chargé d'étudier Les Opérations de Paix Des Nations Unies. A/70/95–S/2015/446*.* United Nations.
Hoffmann, S. (1998). *World Disorders: Troubled Peace in the Post-Cold War Era.* Rowman & Littlefield.
Hoffmann, S. (2006). Raymond Aron et la théorie des relations internationales. *Politique étrangère, 4*, 723–734. https://doi.org/10.3917/pe.064.0723
Holeindre, J.-V., & Vilmer, J.-B. J. (2015). Pour Des War Studies En France: Un Diagnostic et Des Propositions. *Revue Défense Nationale, 785*(10), 53–59. https://doi.org/10.3917/rdna.785.0053
Hopgood, S. (2013). *The Endtimes of Human Rights.* Cornell University Press.
Howard, L. M. (2008). *UN Peacekeeping in Civil Wars.* Cambridge University Press.
Howard, L. (2019a). Peacekeeping Is Not Counterinsurgency. *International Peacekeeping, 26*(5), 545–548. https://doi.org/10.1080/13533312.2019.1677288

Howard, L. (2019b). *Power in Peacekeeping* (1st ed.). Cambridge University Press. https://doi.org/10.1017/9781108557689

Howard, L. M. (2019c). *Power in Peacekeeping*. Cambridge University Press.

Howard, L. M., & Stark, A. (2018). How Civil Wars End: The International System, Norms, and the Role of External Actors. *International Security, 42*(3), 127–171. https://doi.org/10.1162/ISEC_a_00305

Huddy, L., Sears, D. O., & Levy, J. S. (2003). *The Oxford Handbook of Political Psychology* (2nd ed.). Éditeur, Oxford University Press.

Hunt, C. T. (2017). All Necessary Means to What Ends? The Unintended Consequences of the 'Robust Turn' in UN Peace Operations. *International Peacekeeping, 24*(1), 108–131. https://doi.org/10.1080/13533312.2016.1214074

Ignatieff, M. (2003). *Empire Lite: Nation-Building in Bosnia, Kosovo and Afghanistan. A Vintage Original.* Vintage.

Ikenberry, G. J. (2001). *After Victory: Institutions, Strategic Restraint, and the Rebuilding of Order after Major Wars. Princeton Studies in International History and Politics.* Princeton University Press.

Ikenberry, G. J. (2011). *Liberal Leviathan: The Origins, Crisis, and Transformation of the American World Order. Princeton Studies in International History and Politics.* Princeton University Press.

Isoda, V. (2020). Choosing the Flag in the Name of Peace: Why Have the Baltic States (Re)Turned to the United Nations? *International Peacekeeping, Décembre, 11*, 1–33. https://doi.org/10.1080/13533312.2020.1855983

James, A. (1990). *Peacekeeping in International Politics.* Palgrave Macmillan UK. https://doi.org/10.1007/978-1-349-21026-8

Jeangène, V. (2020). *Jean-Baptiste. Théories des relations internationales. Que sais-je?* Presses Universitaires de France. Retrieved from https://www.cairn.info/theories-des-relations-internationales%2D%2D9782130785644.htm

Jervis, R. (2017). *Perception and Misperception in International Politics* (New ed.). Princeton University Press.

Jordaan, E. (2003). The Concept of a Middle Power in International Relations: Distinguishing between Emerging and Traditional Middle Powers. *Politikon, 30*(1), 165–181. https://doi.org/10.1080/0258934032000147282

Kalb, M. G. (1982, Décembre 19). *THE U.N.'S EMBATTLED PEACEKEEPER*. Retrieved from https://www.nytimes.com/1982/12/19/magazine/the-un-s-embattled-peacekeeper.html

Karlsrud, J. (2013). Special Representatives of the Secretary-General as Norm Arbitrators? Understanding Bottom-up Authority in UN Peacekeeping. *Global Governance: A Review of Multilateralism and International Organizations, 19*(4), 525–544. https://doi.org/10.1163/19426720-01904004

Karlsrud, J. (2015). The UN at War: Examining the Consequences of Peace-Enforcement Mandates for the UN Peacekeeping Operations in the CAR, the

DRC and Mali. *Third World Quarterly, 36*(1), 40–54. https://doi.org/10.1080/01436597.2015.976016

Karlsrud, J. (2016). *Norm Change in International Relations: Linked Ecologies in UN Peacekeeping Operations*. Routledge.

Karlsrud, J. (2017). Towards UN Counter-Terrorism Operations? *Third World Quarterly, 38*(6), 1215–1231. https://doi.org/10.1080/01436597.2016.1268907

Karlsrud, J. (2019a). For the Greater Good?: 'Good States' Turning UN Peacekeeping towards Counterterrorism. *International Journal: Canada's Journal of Global Policy Analysis, 74*(1), 65–83. https://doi.org/10.1177/0020702019834725

Karlsrud, J. (2019b). From Liberal Peacebuilding to Stabilization and Counterterrorism. *International Peacekeeping, 26*(1), 1–21. https://doi.org/10.1080/13533312.2018.1502040

Karlsrud, J., & Oksamytna, K. (2019). Norms and Practices in UN Peacekeeping: Evolution and Contestation. *International Peacekeeping, 26*(3), 253–254. https://doi.org/10.1080/13533312.2019.1600637

Karlsrud, J., & Reykers, Y. (2020). Ad Hoc Coalitions and Institutional Exploitation in International Security: Towards a Typology. *Third World Quarterly, 9*, 1–19. https://doi.org/10.1080/01436597.2020.1763171

Kegley, C. W. (1995). *Controversies in International Relations Theory: Realism and the Neoliberal Challenge*. St. Martin's Press [u.a.].

Keohane, R. O. (2011). Global Governance and Legitimacy. *Review of International Political Economy, 18*(1), 99–109. https://doi.org/10.1080/09692290.2011.545222

Koops, Joachim Alexander, and Giulia Tercovich. European Approaches to United Nations Peacekeeping: Towards a Stronger Re-engagement? , 2018.

Koops, J. A., MacQueen, N., Tardy, T., & Williams, P. (Eds.). (2014). *The Oxford Handbook of United Nations Peacekeeping Operations*. Oxford University Press.

La Charte Des Nations Unies. (2014). *La Pensée 377*(1), 141–156. https://doi.org/10.3917/lp.377.0141

Larose, C. (2010). David Ambrosetti, 2009, Normes et rivalités diplomatiques à l'ONU. Le Conseil de sécurité en audience, coll. Regards sur l'international, no 8, Bruxelles, PIE–Peter Lang, 352 p. *Études internationales, 41*(2), 290. https://doi.org/10.7202/044630ar

Leprette, J. (1988). Le Conseil de sécurité et la Résolution 377 A (1950). *Annuaire français de droit international, 34*(1), 424–435. https://doi.org/10.3406/afdi.1988.2847

Lequesne, C. (2018). Chapitre 8. Les États et Leur Outil Diplomatique. In *Manuel de Diplomatie* (pp. 143–161). Presses de Sciences Po. https://doi.org/10.3917/scpo.balza.2018.01.0143

Louis, M., & Maertens, L. (2021). *Why International Organizations Hate Politics: Depoliticizing the World*. Routledge.
Lyall, J., & Wilson, I. (2009). Rage Against the Machines: Explaining Outcomes in Counterinsurgency Wars. *International Organization, 63*(1), 67–106. https://doi.org/10.1017/S0020818309090031
Lyman, P. N. (2004). Ralph Bunche's International Legacy: The Middle East, Congo, and United Nations Peacekeeping. *The Journal of Negro Education, 73*(2), 159. https://doi.org/10.2307/3211272
Mac Ginty, R. (2010). Hybrid Peace: The Interaction Between Top-Down and Bottom-Up Peace. *Security Dialogue, 41*(4), 391–412. https://doi.org/10.1177/0967010610374312
Mac Ginty, R. (2012). Against Stabilization. *Stability: International Journal of Security and Development, 1*(1), 20. https://doi.org/10.5334/sta.ab
MacQueen, Norrie. Peacekeeping and the International System. , 2006.
Maertens, L. (2016). Ouvrir La Boîte Noire: Observation Participante et Organisations Internationales. *Terrains/Théories, 5*. https://doi.org/10.4000/teth.749
Malone, D. M., & Day, A. (2020). The UN at 75: How Today's Challenges Will Shape the Next 25 Years. *Global Governance: A Review of Multilateralism and International Organizations, 26*(2), 236–250. https://doi.org/10.1163/19426720-02602004
Mearsheimer, J. J. (1994). The False Promise of International Institutions. *International Security, 19*(3), 5. https://doi.org/10.2307/2539078
Meijer, H., & Wyss, M. (Eds.). (2018). *The Handbook of European Defence Policies and Armed Forces* (1st ed.). Oxford University Press.
Moravcsik, A. (1997). Taking Preferences Seriously: A Liberal Theory of International Politics. *International Organization, 51*(4), 513–553. https://doi.org/10.1162/002081897550447
Murray, A. J. H. (1996). The Moral Politics of Hans Morgenthau. *The Review of Politics, 58*(1), 81–108. https://doi.org/10.1017/S0034670500051676
Nadin, P. (Ed.). (2018). *The Use of Force in UN Peacekeeping. Routledge Global Institutions Series*. Routledge, Taylor & Francis Group.
Nations Unies. Département des opérations de maintien de la paix, and Nations Unies. Département de l'appui aux missions. (2008). *United Nations Peacekeeping Operations: Principles and Guidelines*. United Nations, Department of Peacekeeping Operations: Department of Field Support.
Neack, L. (1995). UN Peace-Keeping: In the Interest of Community or Self? *Journal of Peace Research, 32*(2), 181–196. https://doi.org/10.1177/0022343395032002005
Novosseloff, A. (2003). *Le Conseil de Sécurité Des Nations Unies et La Maîtrise de La Force Armée: Dialectique Du Politique et Du Militaire En Matière de Paix et*

de Sécurité Internationales. Organisation Internationale et Relations Internationales 56. E. Bruylant.

Novosseloff, A. (2015). In J. A. Koops, T. Tardy, N. MacQueen, & P. D. Williams (Eds.), *United Nations Interim Force in Lebanon (UNIFIL I & II)*. Oxford University Press. https://doi.org/10.1093/oxfordhb/9780199686049.013.68

Novosseloff, A. (2021). *Le Conseil de sécurité des Nations Unies: entre impuissance et toute-puissance* (2nd édition ed.). CNRS Editions, collection Biblis 137.

Nye, J. S. (2004). *Soft Power: The Means to Success in World Politics* (1st ed.). Public Affairs.

Oksamytna, K., & Karlsrud, J. (2020). *United Nations Peace Operations and International Relations Theory*. Manchester Univ. Press.

Oksamytna, K., Bove, V., & Lundgren, M. (2020). Leadership Selection in United Nations Peacekeeping. *International Studies Quarterly, 4*, sqaa023. https://doi.org/10.1093/isq/sqaa023

ONU. (2008). Doctrine Capstone. Opérations de Maintien de La Paix Des Nations Unies. Principes et Orientations.

Paris, R. (2000). Broadening the Study of Peace Operations. *International Studies Review, 2*(3), 27–44. https://doi.org/10.1111/1521-9488.00214

Paris, R. (2003). Peacekeeping and the Constraints of Global Culture. *European Journal of International Relations, 9*(3), 441–473. https://doi.org/10.1177/13540661030093004

Paris, R. (2014). The Geopolitics of Peace Operations: A Research Agenda. *International Peacekeeping, 21*(4), 501–508. https://doi.org/10.1080/13533312.2014.946743

Paris, R. (2020). The Right to Dominate: How Old Ideas About Sovereignty Pose New Challenges for World Order. *International Organization, 23*, 1–37. https://doi.org/10.1017/S0020818320000077

Petiteville, F. (2009). *Le Multilatéralisme*. Montchrestien.

Pouligny, B. (2004). *Ils Nous Avaient Promis La Paix: Opérations de l'ONU et Populations Locales. Mondes*. Presses de la Fondation nationale des sciences politiques.

Pouliot, V. (2017). *L'ordre hiérarchique international: les luttes de rang dans la diplomatie multilatérale. Collection relations internationales 7*. Presses de Sciences Po.

Pouliot, V. (2021). The Gray Area of Institutional Change: How the Security Council Transforms Its Practices on the Fly. *Journal of Global Security Studies, 6*(3), ogaa043. https://doi.org/10.1093/jogss/ogaa043

Pugh, M. (2004). Peacekeeping and Critical Theory. *International Peacekeeping, 11*(1), 39–58. https://doi.org/10.1080/1353331042000228445

Ravenhill, J. (1998). Cycles of Middle Power Activism: Constraint and Choice in Australian and Canadian Foreign Policies. *Australian Journal of International Affairs, 52*(3), 309–327. https://doi.org/10.1080/10357719808445259

Renshon, S., & Welch Larson, D. (Eds.). (2003). *Good Judgment in Foreign Policy: Theory and Application*. Rowman and Littlefield.

Rhodes, B. (n.d.). *The 9/11 Era Is Over The Coronavirus Pandemic and a Chapter of History That Should Have Expired Long Ago*. The Atlantic. Retrieved from https://www.theatlantic.com/ideas/archive/2020/04/its-not-septembre-12-anymore/609502/

Richmond, O. P., & Ginty, R. M. (2015). Where Now for the Critique of the Liberal Peace? *Cooperation and Conflict, 50*(2), 171–189. https://doi.org/10.1177/0010836714545691

Rietjens, S., & Ruffa, C. (2019). Understanding Coherence in UN Peacekeeping: A Conceptual Framework. *International Peacekeeping, 26*(4), 383–407. https://doi.org/10.1080/13533312.2019.1596742

Riggs, R. E. (1978). The United States and Diffusion of Power in the Security Council. *International Studies Quarterly, 22*(4), 513. https://doi.org/10.2307/2600165

Rivlin, B., & Gordenker, L. (Eds.). (1993). *The Challenging Role of the UN Secretary-General: Making "the Most Impossible Job in the World" Possible*. Praeger.

Roberts, A. (1995). From San Francisco to Sarajevo: The UN and the Use of Force. *Survival, 37*(4), 7–28. https://doi.org/10.1080/00396339508442813

Roosevelt, F. (1950). Note du Président Roosevelt le 13 Novembre 1942. In E. Roosevelt (dir.), *FDR: His Personal Letters (1928–1945)* (Vol. II, pp. 1366–1367).

Ruffa, C. (2018). *Military Cultures in Peace and Stability Operations: Afghanistan and Lebanon* (1st ed.). University of Pennsylvania Press.

Sandor, A. (2020). The Power of Rumour(s) in International Interventions: MINUSMA's Management of Mali's Rumour Mill. *International Affairs, 96*(4), 913–934. https://doi.org/10.1093/ia/iiaa092

Schatz, E. (Ed.). (2009). *Political Ethnography: What Immersion Contributes to the Study of Power*. The University of Chicago Press.

Schmitt, O. (Ed.). (2018). *Raymond Aron and International Relations. The New International Relations*. Routledge.

Schweller, R. L. (1994). Bandwagoning for Profit: Bringing the Revisionist State Back In. *International Security, 19*(1), 72–107.

Smouts, M.-C., & Diener, I. (Eds.). (1994). *L'Onu et La Guerre: La Diplomatie En Kaki. Espace International 12*. Editions Complexe.

Spaeth, K. (2001). Book Review: Stanley Hoffmann, World Disorders: Troubled Peace in the Post-Cold War Era (Lanham: Rowman & Littlefield Publishers, 2000, 288 Pp., $22.95 Pbk.). *Millennium: Journal of International Studies, 30*(2), 433–435. https://doi.org/10.1177/03058298010300020425

Spooner, K. A. (2009). *Canada, the Congo Crisis, and UN Peacekeeping, 1960–64*. UBC Press.

Steffek, J. (2015). The Output Legitimacy of International Organizations and the Global Public Interest. *International Theory, 7*(2), 263–293. https://doi.org/10.1017/S1752971915000044

Street, J., & Altiok, A. (2020). *A Fourth Pillar for the United Nations? The Rise of Counter-Terrorism*. SaferWorld.

Sur, S. (2004). Le conseil de sécurité: blocage, renouveau et avenir. *Pouvoirs, 109*(2), 61. https://doi.org/10.3917/pouv.109.0061

Sur, S. (2021). *Relations internationales. 7e édition refondu*. Précis Domat.

Tankel, S. (2020). US Counterterrorism in the Sahel: From Indirect to Direct Intervention. *International Affairs, 96*(4), 875–893. https://doi.org/10.1093/ia/iiaa089

Tardy, T. (2009). Chapitre 3. L'Organisation des Nations Unies: l'éternel recommencement? In *Gestion de crise, maintien et consolidation de la paix. Crisis* (pp. 55–92). De Boeck Supérieur. Retrieved from https://www.cairn.info/gestion-de-crise-maintien-et-consolidation%2D%2D9782804116392-p-55.htm

Tardy, T. (2015). In J. A. Koops, T. Tardy, N. MacQueen, & P. D. Williams (Eds.), *United Nations Protection Force (UNPROFOR—Bosnia and Herzegovina)*. Oxford University Press. https://doi.org/10.1093/oxfordhb/9780199686049.013.37

Teurtrie, D. (2021). *Russie: le retour de la puissance. Objectif Monde*. Armand Colin.

Thant, M.-U., & Scott, A. (2007). *The UN Secretariat: A Brief History (1945–2006)*. International Peace Academy.

Thurston, A. (2020). *Jihadists of North Africa and the Sahel: Local Politics and Rebel Groups* (1st ed.). Cambridge University Press. https://doi.org/10.1017/9781108771160

Urban, A. (2006). Les États-Unis et l'ONU dans les années 1950. *Relations internationales, 127*(3), 47. https://doi.org/10.3917/ri.127.0047

Urquhart, B. E. (1987). *A Life in Peace and War*. Weidenfeld and Nicolson.

Urquhart, B. (1993). *Ralph Bunche: An American Life* (1st ed.). W.W. Norton.

Utley, Rachel E., ed. Major Powers and Peacekeeping: Perspectives, Priorities and the Challenges of Military Intervention. , 2006.

Vaïsse, M. (2009). *La Puissance Ou l'influence? La France Dans Le Monde Depuis 1958*. Fayard.

Van Langenhove, F. (1960). La Crise Du Système de Sécurité Collective Des Nations Unies 1946–1957. By Fernand Van Langenhove. Brussels: Institut Royal Des Relations Internationales; The Hague: Martinus Nijhoff, 1958. Pp. 272. Index. Fr. 250. *American Journal of International Law, 54*(1), 216. https://doi.org/10.2307/2195132

van Willigen, N. (2016). A Dutch Return to UN Peacekeeping? *International Peacekeeping, 23*(5), 702–720. https://doi.org/10.1080/13533312.2016.1235095

Vanlangenhove, F. (1970). La sécurité collective des Nations Unies au vingt-cinquième anniversaire de la Charte. *Bulletin de la Classe des lettres et des sciences morales et politiques, 56*(1), 545–561. https://doi.org/10.3406/barb.1970.54958

Védrine, H., & Boniface, P. (2001). *1991–2001: dix années qui ébranlèrent le monde*. IRIS: PUF.

Weinlich, S. (2014). *The UN Secretariat's Influence on the Evolution of Peacekeeping*. Palgrave Macmillan.

Wendt, Alexander. Social Theory of International Politics, 1999.

Williams, P. D. (2015). In J. A. Koops, T. Tardy, N. MacQueen, & P. D. Williams (Eds.), *United Nations Operation in Somalia II (UNOSOM II)*. Oxford University Press. https://doi.org/10.1093/oxfordhb/9780199686049.013.41

Williams, P. D. (2016). AMISOM under Review. *The RUSI Journal, 161*(1), 40–49. https://doi.org/10.1080/03071847.2016.1152120

Williams, P. D., & Bellamy, A. J. (2021). *Understanding Peacekeeping* (3rd ed.). Polity Press.

Williams, P. D., & Boutellis, A. (2014). Partnership Peacekeeping: Challenges and Opportunities in the United Nations-African Union Relationship. *African Affairs, 113*(451), 254–278. https://doi.org/10.1093/afraf/adu021

Wiseman, H. (Ed.). (1983). *Peacekeeping, Appraisals & Proposals*. Pergamon Press.

Yin, R. K. (2014). *Case Study Research: Design and Methods* (5th ed.). Sage.

Zacher, M. W. (1970). *Dag Hammarskjold's United Nations. Columbia University Studies in International Organization*. Columbia University Press.

Index[1]

A

B

[1] Note: Page numbers followed by 'n' refer to notes.

A. Boutellis, *The UN Stabilization Mission in Mali (MINUSMA)*, Twenty-first Century Perspectives on War, Peace, and Human Conflict, https://doi.org/10.1007/978-3-031-66466-3

The manufacturer's authorised representative in the EU is Springer Nature Customer Service Centre GmbH, Europaplatz 3, 69115 Heidelberg, Germany. If you have any concerns regarding our products, please contact ProductSafety@springernature.com

Printed and bound by CPI Group (UK) Ltd, Croydon, CR0 4YY
01/10/2026
02225912-0006